W0113866

Mathematical Meditations

Mathematical Meditations identifies, explores, and celebrates those aspects of mathematics that are good for you and your overall wellbeing. It is necessary for everyone to have a little time to think every so often: to contemplate, meditate, and try to understand where you are and what is going on around you. Mathematics can help you with all of that.

The Meditations in this book are the product of thousands of years of mathematical discourse. As you read through the book and work through the various exercises, you will discover new mechanisms that allow you to contemplate and understand some complex mathematical principles. However, the focus will always be wider than a mere dry comprehension of theory, as you will be encouraged to meditate upon the deeper intrinsic beauty of mathematics and what it can reveal to us about the world around us.

Features

- An original, engaging narrative format replete with novel exercises and examples.
- Could be used in a classroom setting for liberal arts students, mathematics undergraduates, or high school teachers.
- Accessible to anyone who wants to explore a different kind of perspective on mathematics.

AK Peters/CRC Recreational Mathematics Series

Series Editors

Robert Fathauer
Snezana Lawrence
Jun Mitani
Colm Mulcahy
Peter Winkler
Carolyn Yackel

Mathematical Puzzles
Revised Edition
Peter Winkler

Mathematics of Tabletop Games
Aaron Montgomery

Puzzle and Proof
A Decade of Problems from the Utah Math Olympiad
Samuel Dittmer, Hiram Golze, Grant Molnar, and Caleb Stanford

A Stitch in Line
Mathematics and One-Stitch Sashiko
Katherine Seaton

The Four Corners of Mathematics
A Brief History, from Pythagoras to Perelman
Thomas Waters

Intermediate Poker Mathematics
Mark Bollman

Mathematical Meditations
Snezana Lawrence

For more information about this series please visit: www.routledge.com/AK-PetersCRC-Recreational-Mathematics-Series/book-series/RECMATH?pd=published,forthcoming&pg=2&pp=12&so=pub&view=list

Mathematical Meditations

Snezana Lawrence

CRC Press
Taylor & Francis Group
Boca Raton   London   New York

CRC Press is an imprint of the
Taylor & Francis Group, an **informa** business

AN A K PETERS BOOK

Designed cover image: Shutterstock

First edition published 2025
by CRC Press
2385 NW Executive Center Drive, Suite 320, Boca Raton FL 33431

and by CRC Press
4 Park Square, Milton Park, Abingdon, Oxon, OX14 4RN

CRC Press is an imprint of Taylor & Francis Group, LLC

© 2025 Snezana Lawrence

Reasonable efforts have been made to publish reliable data and information, but the author and publisher cannot assume responsibility for the validity of all materials or the consequences of their use. The authors and publishers have attempted to trace the copyright holders of all material reproduced in this publication and apologize to copyright holders if permission to publish in this form has not been obtained. If any copyright material has not been acknowledged please write and let us know so we may rectify in any future reprint.

Except as permitted under U.S. Copyright Law, no part of this book may be reprinted, reproduced, transmitted, or utilized in any form by any electronic, mechanical, or other means, now known or hereafter invented, including photocopying, microfilming, and recording, or in any information storage or retrieval system, without written permission from the publishers.

For permission to photocopy or use material electronically from this work, access *www.copyright.com* or contact the Copyright Clearance Center, Inc. (CCC), 222 Rosewood Drive, Danvers, MA 01923, 978-750-8400. For works that are not available on CCC please contact *mpkbookspermissions@tandf.co.uk*

Trademark notice: Product or corporate names may be trademarks or registered trademarks and are used only for identification and explanation without intent to infringe.

Library of Congress Cataloging-in-Publication Data
Names: Lawrence, Snezana, author.
Title: Mathematical meditations / Snezana Lawrence.
Description: First edition. | Boca Raton : AK Peters/CRC Press, 2025. | Series: AK Peters/
    CRC recreational mathematics series | Includes bibliographical references and index.
Identifiers: LCCN 2024043820 (print) | LCCN 2024043821 (ebook) | ISBN 9781032249070
    (hardback) | ISBN 9781032249063 (paperback) | ISBN 9781003280668 (ebook)
Subjects: LCSH: Mathematical recreations.
Classification: LCC QA95 .L397 2025 (print) | LCC QA95 (ebook) | DDC 510—dc23/eng/20241205
LC record available at https://lccn.loc.gov/2024043820
LC ebook record available at https://lccn.loc.gov/2024043821

ISBN: 978-1-032-24907-0 (hbk)
ISBN: 978-1-032-24906-3 (pbk)
ISBN: 978-1-003-28066-8 (ebk)

DOI: 10.1201/9781003280668

Typeset in Minion Pro
by Apex CoVantage, LLC

*This book is for my Natalija
who has reminded me of the importance
of meditation*

———————————————

Contents

About the Author

 **Snezana Lawrence** is a mathematical historian and mathematics educator and has been working in higher education for the past 15 years. Prior to that, Snezana was a teacher and teacher trainer in the English school system. She was a lead consultant for the Prince's Teaching Institute for Mathematics 2009–2018. Currently Snezana works as a senior lecturer in the Department of Mathematics and Design Engineering at Middlesex University, London, teaching mathematics, mathematics history, and risk.

Snezana's plan for the next few years is to write more popular history of mathematics books and to work more with teachers and people interested in identifying their local history of mathematics in libraries and museums where they live, across Europe. She was the Chair of the History and Pedagogy of Mathematics International Study Group for 2020–2024.

Figures

Introduction

ONCE UPON A TIME, AND NOT, IT SEEMS, SUCH A LONG TIME AGO, THERE WAS A war in my country. I came to England. A friend offered to take me on a walk with the curious group of people who call themselves *ramblers*. In England there is a law that says that if you don't walk on country paths for a year, a farmer who owns that piece of land can close the path. Ramblers keep such paths open. They plan which paths they wish to cover every year, and they do so systematically. Usually, a local authority helps them by recording the paths and the ramblers' walks and negotiating with farmers if any trouble should arise at any point.

So, this was a new thing for me, and a friend, who somehow knew my father (as he did some voluntary work for an international organisation many years previously), said it would be good for me. He suggested that walking on these paths, I will start *claiming* the land as my own. I will begin feeling at home. He also said that all the people in this group of ramblers were rather nice and so I should join to meet some people in my new place of abode.

He was right. Once, we went on one of these walks. There was a lot of mud on the pathway, but we soldiered on. Someone stared singing "mud, mud, glorious mud, there's nothing quite like it to cool the blood!" "What a wonderful tune and lyrics", I exclaimed. I learnt that little song by heart.

Years later, I started a PhD in the history of mathematics. At one of the conferences, I spoke to Bob Burn, a wonderful mathematician and mathematics teacher, who asked me how and why I was interested in the history of mathematics. By this time, I was happy to say that, apart from being reasonably good at mathematics (and its history), I found that this subdiscipline of both mathematics and history was a safe haven for me intellectually. It offered me enough of storytelling and enough of mathematics that allowed me to always keep going. It was also a way out of the turmoil. Above all, it made me feel at home with mathematics, a universal truth-churning pursuit. This was helpful on occasions when I felt unable to make up my mind about politics or personal relationships.

Bob exclaimed: "maths, maths, glorious maths, there's nothing quite like it for cooling the blood!" What a happy coincidence this was. It captured in a line exactly what I was trying to say to Bob. But it gave me another meaning that kept bothering me all these years since: one has to keep going, keep the pathways clear, even if mud is involved. Keep the blood cool. The history of mathematics is a great way of rambling through the fields of both history and mathematics. I will be just your guide here and hope you will continue meditating on various things we mention and come up with your own things later on.

What things, will depend on how mathematically minded you are. You may continue by doing some mathematics, or some further historical research. Or you may keep doodling. There are lots of my own doodles to help you visualise things I write about here. You may search for some novels or musical pieces mentioned throughout, and I have suggested a few along the way.

The book is organised into 12 chapters. The chapters are organised by starting from a subject which is set in its historical context or perspective and then we go on from there. The trail goes on, not necessarily in the way everyone would find natural, but there is a bit of association game going on here. The details of the stories are such that I hope many people will be able to enjoy this narrative, be they familiar with mathematics or not, and learn a little bit of history of it too. I have stripped the book from all technical detail to offer you starters for your own further thinking and meditations.

I have no advice on how to conduct your meditations. How to sit or breathe or where your mind should wander is up to you. Equally, perhaps, there is no prior knowledge of higher mathematics needed for the reading this book. I will introduce some concepts of higher mathematics inevitably and will offer some explanations in the main text. There will be suggestions for further reading or exploration for those who want to learn more at the end of each chapter, and not always relating to mathematics, but in some way connected to what has been said in that chapter.

Is this a book therefore aimed at non-mathematicians? It is not meant to be prescriptive in that way at all. I hope those who know both some and no mathematics at all will be able to enjoy the stories presented here. I hope that people who think mathematics may be beautiful but couldn't access its beauty in a visual way would also be able to find some starting points here. Perhaps you'll get into new *fields* you can ramble through and visit some pathways you may want to keep open for future visits. Remember also that if there's a bit of mud there, getting through it may well be worth your while.

Post Scriptum to Introduction

The mathematicians and others, like artists, architects, writers, and even film directors, will be mentioned throughout. There are a few pages about most mathematicians mentioned at the end of the book. This is done to avoid making stops all the time to include such data in the main text.

The suggested further exploration is given for each chapter, but there is no final bibliography.

Further Exploration

Well, just to introduce you to the format, here's a *further exploration* section for this introduction already. This is how it looks.

I would recommend finding that song that contains the refrain "mud, mud, glorious mud". It is called *Hippopotamus Song*, and was written by Michal Flanders and Donald Swan in 1957.

Bob Burn published a wonderful book *Numbers and Functions, Steps Into Analysis*, with Cambridge University Press, in 1993. There have been two further editions, they are all great, if you are learning about numbers, high time you see it.

CHAPTER 1

Labyrinth

From Minotaur to Chartres

T HE OPENING OF THIS book is about walking and meditation. This is not a new thing or something invented recently. Walking within a covered space is a little bit different. There is usually not so much space as there is if you walk in the open. But you can overcome that by making your path a little longer. And what could be better than making an elaborate labyrinth? If you have ever been to the mediaeval cathedral of Chartres, you would have seen the labyrinth on its floor. It is central to the cathedral. The building works there began in 1145 and were completed in the 13th century. This labyrinth is famous for its amazingly beautiful stained-glass windows that have almost miraculously survived the ravages of time to the present day.

It is thought that walking of the labyrinth in Chartres' Cathedral was done as a penitence or for meditation – or both. We would go too far to make an absolute generalisation on this, but most surviving labyrinths were made in Europe in the Middle Ages, in the Gothic churches and cathedrals. The 19th century, with its Gothic Revival of course meant that more new labyrinths were made across Europe, but Chartres' labyrinth remains one of the greatest to inspire labyrinth enthusiasts.

Whether the meditation reference is true or not, we can certainly imagine that verses from the Bible were used for meditations in cathedrals' labyrinths. It is suggested that a verse could be the one from Isaiah, the 23rd book (Old Testament):

your teachers will be hidden no more; with your own eyes you will see them. Whether you turn to the right or to the left. Your ears will hear a voice behind you, saying: *this is the way, walk it.*

(ISIAH 30:20)

The transience of the context – whoever or whatever we want to quote or ignore as it may be – may have different meaning for each one of us. This will also change at different times of our lives. Such a quote would for some resonate with their experience of some kind of suffering, or on the opposite of that experience, with a bliss that may blind us to the problems we see ahead. We may think of the riches we experienced or witnessed, or a period of hardship. All of the extreme associations that trouble us can be transcended by an act of concentration and by focusing on mathematics. Taking the time to meditate is giving yourself time to make your mind up. Centre yourself, walk the labyrinth, and try to find, while walking it, a sense of doing the right thing.

On the other hand, a labyrinth can be a structure meant to mislead and slow down those who pursue us. It can be seen as a structure that makes us get lost on purpose. What that purpose is, and who is pursuing us, will be different to us all. Maybe you'd like to go into a labyrinth to forget about your daily chores.

A labyrinth has a beginning and an end. There is a starting point to it. That point should be easily identifiable. It is an entrance, an invitation to begin this journey of getting lost in it, or finding a quiet space to hide. Once you reach the end of the labyrinth you can decide what is it you went in for – rest or hide? or exit and rejoin the world refreshed.

Here's the image of the Chartres' labyrinth. Identify some points of interest to get started. Eventually we will trace a path through it.

You will see the little incisions on the outside circle. They may be just enough to mark places of people standing on the outside turned in, out, or around the circle one behind another – you decide. See the centre of this labyrinth too. There are six places just the size of a person to stand in. And the central point is where the one who enters the labyrinth will stay before turning around and going back out.

FIGURE 1   Chartres' labyrinth recreated.

1.1 THE CRETAN LABYRINTH

The Cretan labyrinth is made after a story. The story is from the Greek mythology: King Minos wanted to save his daughter Ariadne from the Minotaur. The Minotaur was a half-man half-beast creature. Ariadne was given a thread. When she led the Minotaur into the labyrinth, she could easily find her way back, but he could not.

The story doesn't make much sense actually. In this labyrinth there is one way in and the same way out. To make the labyrinth itself is not however that easy unless someone shows you how.

So we will do that. Start with a simple square. A square is a wonderful thing. Four equal sides meeting two at each vertex and forming four right angles. One can meditate upon that for a long time. What could you do with a square? How would you make it into a three-dimensional equivalent, a cube? That's something to think about. Here we'll stick with the square at the start of learning how to construct a labyrinth.

A simple square it is.

FIGURE 2   A simple square.

Proceed dividing this square into four squares.

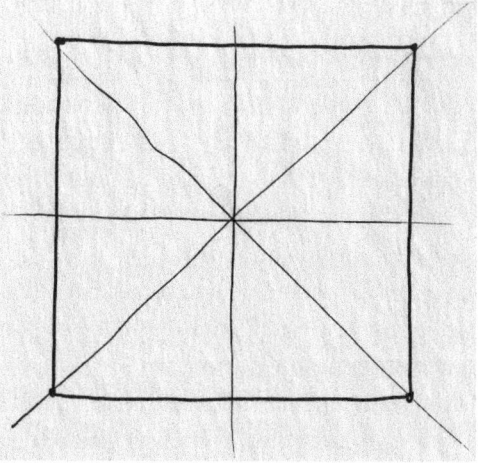

FIGURE 3   Our square divided into quarters.

Each quarter of a square is divided further into quarters by its own diagonals.

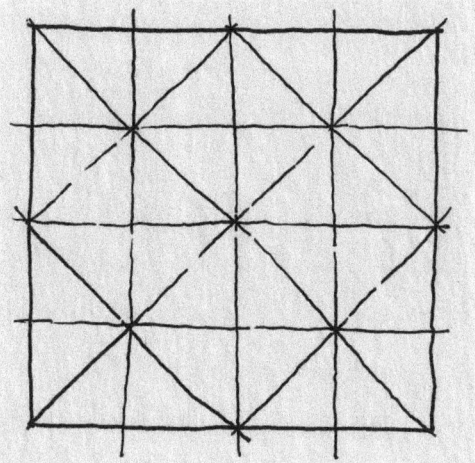

FIGURE 4   Quarters divided into further quarters.

In outer squares draw quarters of a circle towards the inside of the square.

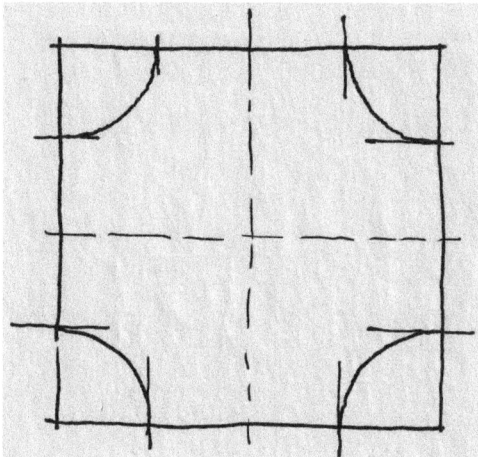

FIGURE 5   Square with rounded edges (the other way around).

Let's now look at what we have. There's an underlying structure beginning to show. First notice a cross in the centre of the square. Identify all the end points of all the lines that meet with the perimeter of the original, largest square. These will determine your 'seed' to start structuring a labyrinth. Number these as follows.

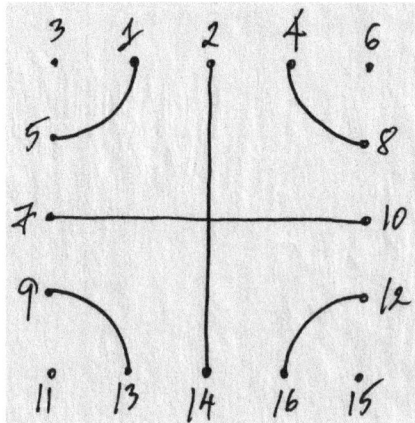

FIGURE 6   The seed of our Cretan labyrinth.

A seed of a Cretan labyrinth is generated, consisting of the quarter circles in the corners and a pair of crossed divisions of the square.

Now connect the points: 1 and 2, 3 and 4, 5 and 6, and so on. The seeds of the Cretan labyrinth begin to generate the labyrinth lines by connecting certain points.

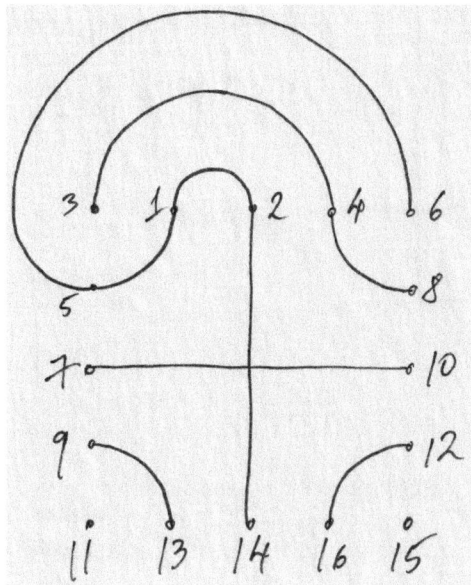

FIGURE 7   Seed germinates.

You now have the beginning of a structure of a labyrinth, that looks a little like the labyrinth of Chartres. It is not really – this one is easier. Join the points in turn: from 1 to 2, from 2 to 3, from 3 to 4 until you complete all the numbers (15 to 16 is the last bit). The finished Cretan labyrinth should look like this.

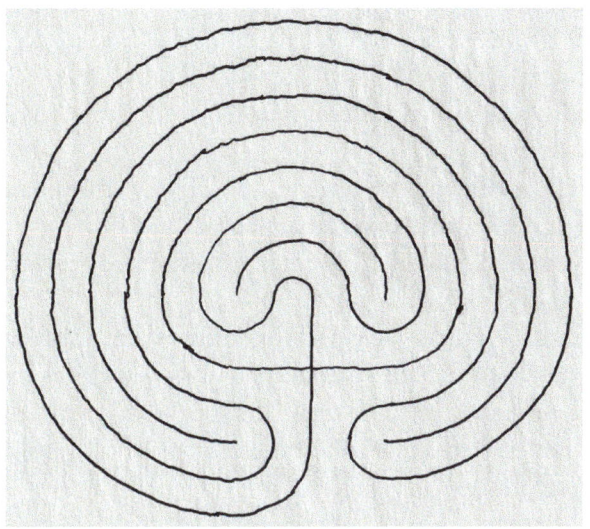

FIGURE 8   A fully grown Cretan labyrinth.

If you walked around it (or traced your steps imagining walking through it) you will get a map of your exploration. Let's now trace of a pathway through the Cretan labyrinth in red colour.

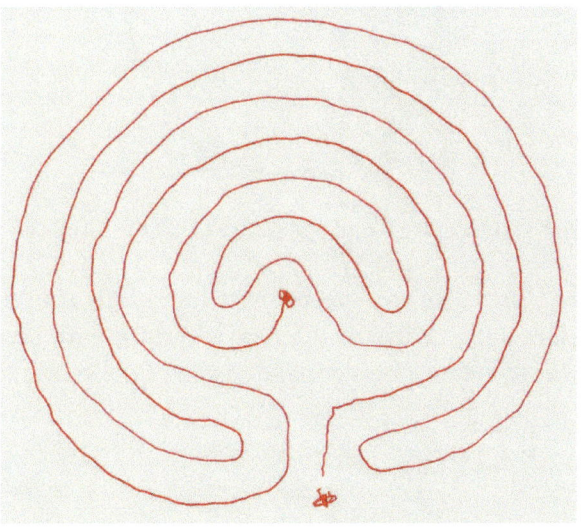

FIGURE 9   A map of going through the Cretan labyrinth.

1.2 THE LABYRINTH OF THE NOTRE DAME DE CHARTRES

Having mastered one way of making labyrinths, let us now turn to see how to construct Chartres' labyrinth. This one has more turns and a more complicated structure. It is believed that it was built sometime between 1201 and 1205. Not all labyrinths have a *seed* like the previous one. We'll investigate one which doesn't have the same kind of beginning and ends in a very different way from the previous labyrinth we constructed.

We will start by drawing first the inner *chamber*. This has a space for six people. Depending on whether you are actually making a large or a small doodle in your notebook, this will determine the size of your eventual finished model or drawing.

Builders and architects have usually used 60 cm (or a little less than 24 in) to be the width of a person. If that person's width is measured sideways then the size should be 30 cm. This means that you can use a circle of that diameter to determine one person's position in space. From here we start.

A circle of diameter of 60 cm is the centre of our labyrinth. A centre of a new labyrinth begins from a circle in which one would stand.

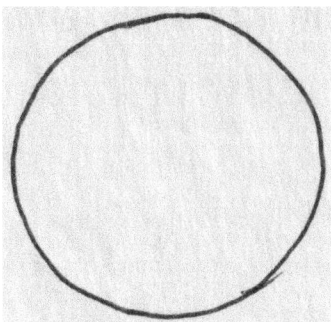

FIGURE 10   You at the centre – and the centre of a labyrinth is your size.

Around that circle, we will create space for exactly six semicircles surrounding it. Draw the centre of the new labyrinth with six circles intertwined in a way to create a hexagonal division of the central circle.

FIGURE 11   Make space for others around you.

This will determine the inner sanctum of the labyrinth.

Let's keep that 60 cm as a space that will be needed for a person to walk through the labyrinth. In our structure we don't include the width of a wall or any external structure that could be built around the pathways.

Chartres has 11 concentric circles within which its structure is confined. Let us then do that next. Next we draw the outline of the Chartres labyrinth with the central figure of six intertwined circles and further concentric circles around it.

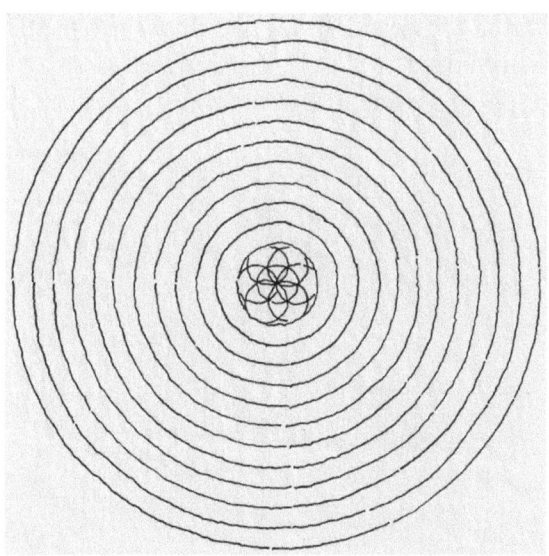

FIGURE 12   The outline of our big Chartres-like labyrinth.

From here we want to divide the whole into four quarters. Let us number the concentric circles, eleven in total.

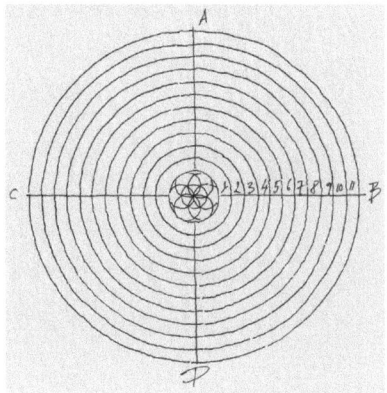

FIGURE 13   Division of our outline into quarters and concentric circles, numbered.

Using letters, let's start by naming the lines that divide our structure into quarters as *A*, *B*, *C*, and *D*. Start from *A*, go to *B*, then *C* skipping the *D*. Finally label the lower bar dividing the circle into four as *D*. The reason for this ordering is that we will come to *D* at the end of our construction. Now look at the intersections of circles with the dividing lines. At *A* we will connect three first concentric circles along the line *AD*, going from the inner circle, and make a break (of one circle) every time we do that. For *B* we will do the same but start *removed* two circles (from outside), and from *C* the same also, removed one circle. We won't touch *D* yet, that will come last. We will get the following diagram.

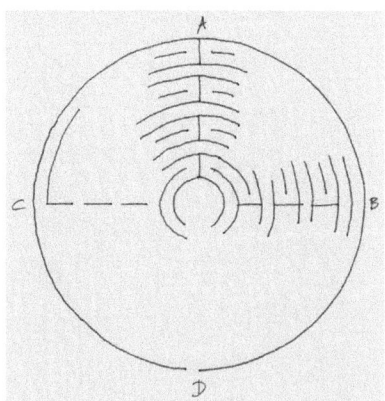

FIGURE 14   Slowly setting the boundaries between turns in our path.

We are well on our way of constructing this labyrinth, only a few more steps are left. First see where the connections between the circles are on the *AD* and *BC* axes, and note that these will be obstacles – walls, boundaries, call them as you wish.

All the middle circles leading to these obstacles will need to stop a little short before the end of their quarter mark to allow for the movement around the side wall. Side walls are straight. It's easier if you look at the next diagram. Pay attention to the new points *Q* and *P*, they show you how and why that is done.

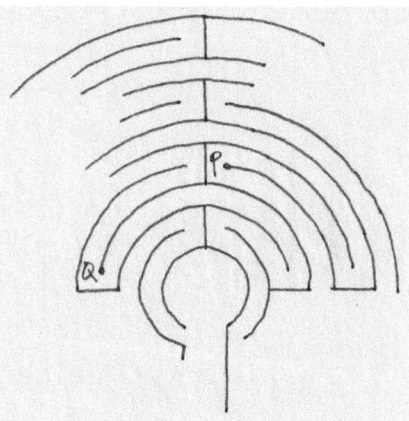

FIGURE 15   Leaving space for movement.

To finish this part of the work, let us do all the circular walls that are in the upper quarter-circle left to *A*. We are now constructing fully the diagram of one quarter side of the labyrinth.

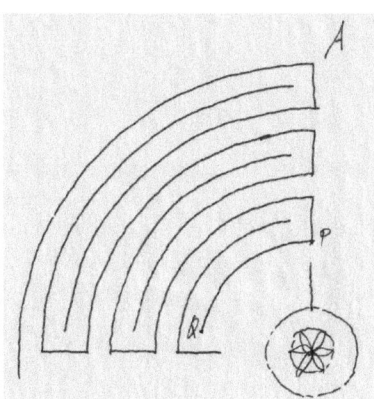

FIGURE 16   One quarter circle finished.

We could continue in the same fashion, complete all the other circular shapes between the points you have identified either as the end wall points or the points on the arcs that stop short of the wall (to allow the movement through the labyrinth). But it will be easier if we knew first where we should get into the labyrinth and how we can get out of it.

You can check what goes on in the real Chartres. The entrance and exit are next to each other, around D. So let us construct that first. The first of the circles (the innermost) and the last circle are here joined by a line/wall that is positioned on the middle axis.

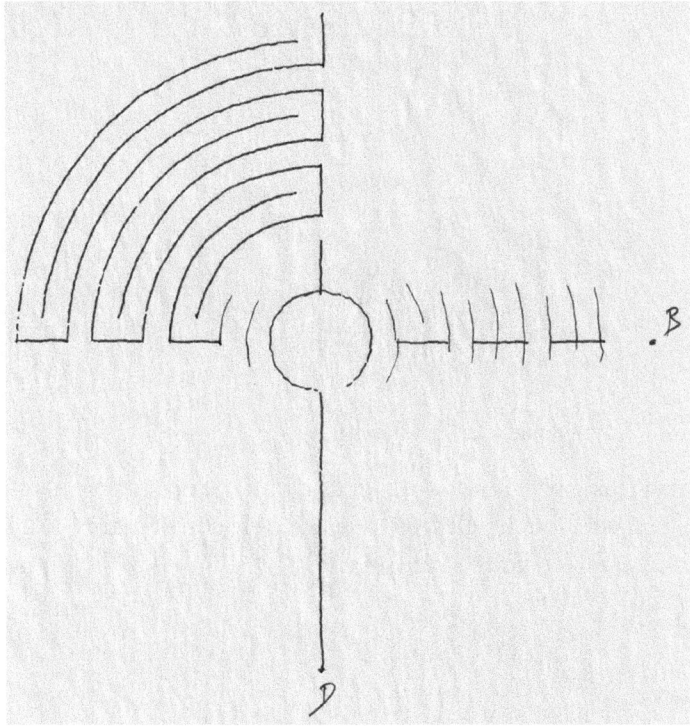

FIGURE 17   Noting the entrance with a straight aim at the centre.

Let us now number our circles from centre towards D. At this point we look at the axis from centre to D – there are two crucial points here, and they are now labelled as X and Y. They show us that, on the right-hand side of the central line, there will be a straight line that will close off some paths, and on the left the straight line (wall, obstruction) that will shut some circular paths too. We construct the entrance to the labyrinth by numbering the concentric circles and working out where the entrance will be in the middle of the concentric circle, labelled 6.

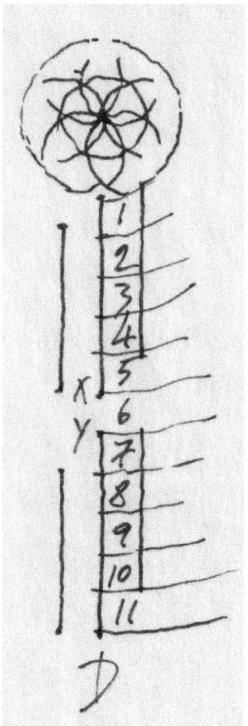

FIGURE 18   Noting where the entrance path and branches are.

So let's use those points *X* and *Y* and connect them to wherever we think they can go upwards towards *C* and *B* whilst leaving space for the movement within our path. We finalise the branches from the centre of the entrance by working out where the boundaries of pathways are.

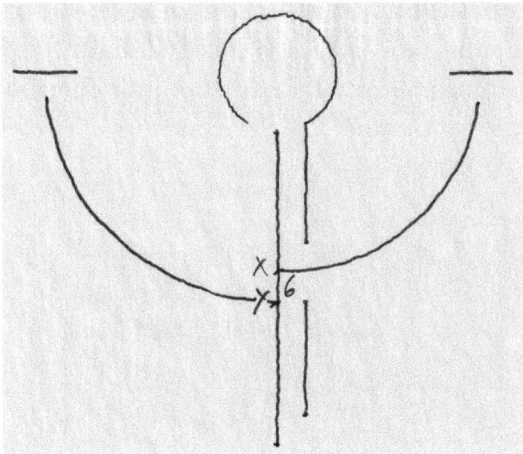

FIGURE 19   Finalising the branches from the centre.

Connect all the points you have identified using the same kind of think-ing we used in the previous few pages, and you will end up with your Chartres' Cathedral labyrinth.

FIGURE 20   Finalising our great labyrinth of Chartres.

Making mistakes while doodling your labyrinth is ok – every mistake will make you think where you went wrong, and, like using a thread of

Ariadne, you will walk a little bit back and perhaps try doing it right the next time. When you finish your labyrinth, try walking through it in your imagination – perhaps draw a red line of your path, your own thread, which you can use to get out.

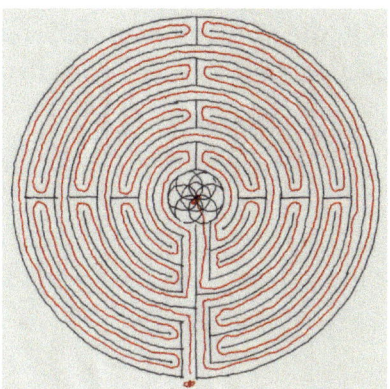

FIGURE 21   Walking through the great labyrinth.

For me the most enjoyable part of doing something like this is complete concentration needed to complete the structure. And there is of course an additional benefit. By practicing this kind of meditative practice, you learn how to see the underlying structures of the things around you. If you have completed your drawing, you will draw double pleasure of knowing the structure, and *walking* the labyrinthine path. The pathway itself tells a story, a trace of your completed task.

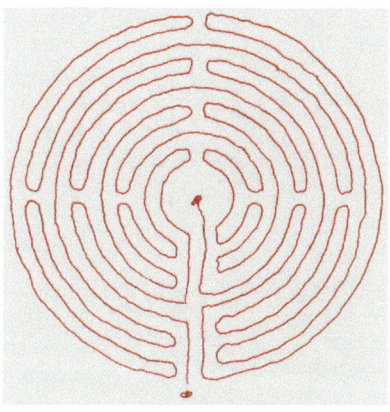

FIGURE 22   Pathway through the great labyrinth.

1.3 FURTHER EXPLORATION

There's a book still going on strong – it was published in 1922 by William Matthews, *Mazes and Labyrinths: A General Account of Their History and Developments*; London: Longmans, Green, and co. It comes highly recommended.

If you haven't seen it, this is the time to find some time to watch it – *Labyrinth*, a 1986 movie starring, among others, the late David Bowie.

CHAPTER 2

Pathways and Bridges

NOT ALL PATHWAYS ARE carefully designed as those we make while walking the labyrinths. You know, for example, that from your home to your favourite coffee shop you can take at least three different paths. Perhaps you don't have a favourite coffee shop, or even three different pathways that can lead you to it, but you get my gist. We all have favourite paths in our everyday life.

2.1 THE FAMOUS SEVEN BRIDGES

Recently a legend has emerged that Immanuel Kant (1724–1804) who was born and lived all of his life in the city of Königsberg, apparently never leaving it until his death (bar once), came up with the problem concerning the seven bridges that cross the river Pregel. Königsberg (now Kaliningrad), is a city in Prussia (now Russia) which sits on the two sides of the river Pregel. The river branches off in the town centre to form two islands – Keniphof and Lomse. The bridges that connect the islands with the rest of the city can be walked in many ways, and the question arose in the early 18th century of whether there is a pathway that could connect the seven bridges in such a way that each bridge would be crossed only once and all of them would be crossed eventually. The problem was considered by Leonhard Euler (1707–1783), one of the most prolific and inventive mathematicians of all time. Euler was Swiss by birth but at the time when he first learnt of the problem he was already living in St Petersburg where he had moved to in 1727.

Now back to the legend. Kant would have been 12 to have devised the problem – a very unlikely possibility – although he may have had the experience of walking the seven bridges by this age. It is unlikely he would have corresponded with the mayor of the town though. And there is this to consider too: in his own words, years later, Kant said,

> But though all our knowledge begins with experience, it does not follow that it all arises out of experience.

But someone did formulate the problem, and someone must have asked Euler to have a look at it. It may have been Christian Goldbach (1690–1764), who we will meet later, and who came to St Petersburg from Königsberg and became Euler's best friend. In any case, Euler corresponded with the mayor of Danzig (now Gdansk), some 80 miles west of Königsberg. The mayor, Carl Leonhard Gottlieb Ehler (1685–1753), seems to have been the intermediary for the local professor of mathematics, Heinrich Kühn (1690–1769). The letter from the mayor to Euler dated 9 March 1736 says that they had already discussed the problem of the bridges. Euler didn't particularly like it and considered it non-mathematical. This is just so that the reader can understand how something that became central to our sense of mathematical sciences now, was in the 18th century considered very far from it. Euler suggested that the solution to this problem bears little relationship to mathematics. Why would anyone ask a mathematician to solve it then? But well, ok, Euler was happy to give a go and try and solve it.

In this letter Euler used a term the *geometry of position*, which was in fact already used by Leibniz (1646–1716). Leibniz considered it in 1679 as *geometria situs* or *analysis situs*, referring to the study of topics one would today consider to be topological. Topological is geometrical in nature but where metrical facts such as distance, length, or angle play no role. Instead, one concentrates on the nature of geometrical objects and their relationships to each other.

How did Euler solve this problem? Is there a pathway that could connect the seven bridges in such a way that each bridge would be crossed only once and all of them would be crossed eventually? Euler wrote a paper in which he described firstly that perhaps one could, if one wanted, try all the possible ways one can cross the bridges. This he dismissed out of hand as too laborious. Secondly, he simplified the situation he was looking at, his *geometria situs*, and drew diagrams of various cases for similar problems. He drew a diagram that showed the case of the seven bridges with land masses labelled with capital letters and bridges by small letters.

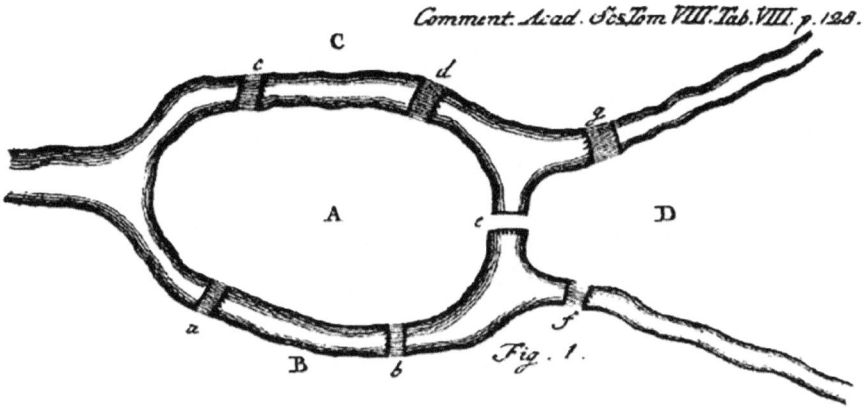

Comment. Acad. Sc. Tom VIII. Tab. VIII. p. 128.

FIGURE 23   Diagram from Euler's 1736 paper *Solutio problematis ad geometriam situs pertinentis* showing the topology of the seven bridges in Königsberg.

Through this way of labelling the diagram he was able to make an abstract system that he could then manipulate. This is a useful skill to learn from the old masters of mathematics, core of which was to *simplify* things.

Let us now start our journey from one land mass and try to cross as many bridges as we can in one go. But remember that you can't come back on the same bridge – you can only cross it once. We will make a doodle showing a try to cross all the seven bridges once and only once.

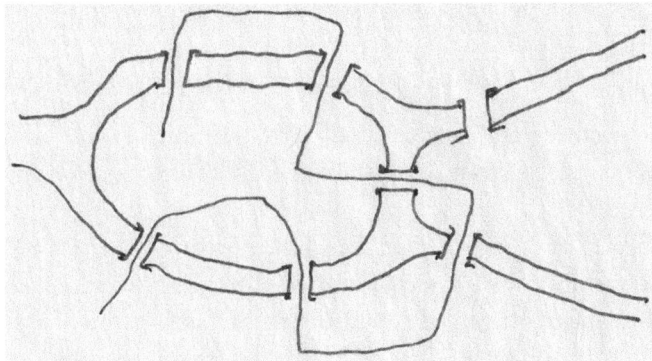

FIGURE 24   The first attempt at covering all the bridges of Königsberg.

That's fine for the moment, but what if we start from another landmass, whichever really, just not the same like the last time? We get a second doodle showing a try to cross all the seven bridges once and only once.

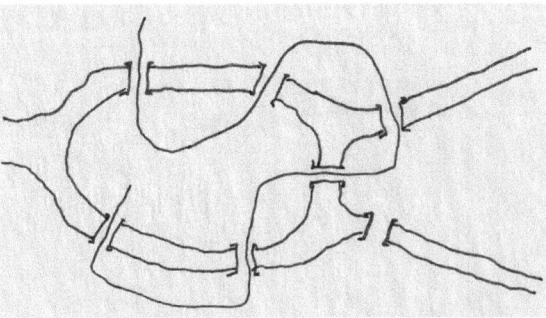

FIGURE 25   Our second attempt at crossing all the bridges once and only once.

We get a very similar diagram the second time around. Let's try once more. A third doodle will be showing a new way of trying to cross all the seven bridges once and only once.

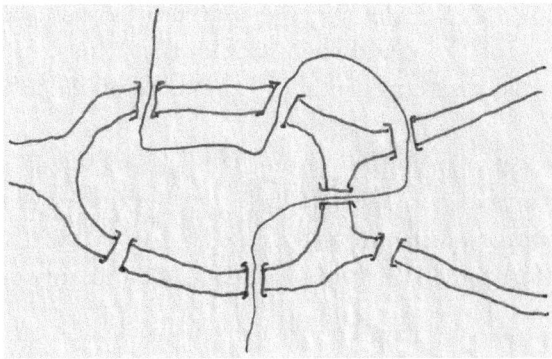

FIGURE 26   The third and final attempt at solving the Königsberg puzzle.

It doesn't matter really where we start, the number of paths out of A, the central island, is 5, but the number of times one has to transverse that landmass is 3.

In general Euler found that the number of bridges to a landmass will be the number of times one will go through land mass +1 and total divided by 2 *if* one wants to cover all the bridges. In our case, for landmass A the number of bridges is 5. Add to that 1 and divide by 2, and you get $\dfrac{5+1}{2} = 3$.

What happens with the landmasses B, C, and D? How many times do we have to go to and from them in order to cover all the bridges? The number of bridges going to B is 3, so if we apply our formula, it will be $\dfrac{3+1}{2} = 2$.

In whichever way you try to cross the bridges to or from *B*, you will transverse that landmass twice. *C* and *D* are the same case as *B*, so however you try, you will transverse lands labelled *C* or *D* twice.

So far so good, the genial power of the method has not yet been revealed to us or probably to Euler at this point. Let's search a little more for the solution.

Let's say we label our journey *AB* when we cross from *A* to *B*, without thinking which one of the bridges the passenger in fact took. There are two bridges, but at this stage let's not make a big deal whether the passenger took one bridge or another to go from one land mass to the next, they just used a bridge to cross the river. That settled, let's say we will then cross from *B* to *D*. The whole journey so far we can now represent as *ABD*. The number of bridges in total one would have had to cross to go from *A* to *B*, and then *D* would be 2. The three-letter journey (*ABD*) would cover two bridges. The next diagram will show us how one can go from the island labelled A, to south of the city, labelled B, and east of the city, labelled C.

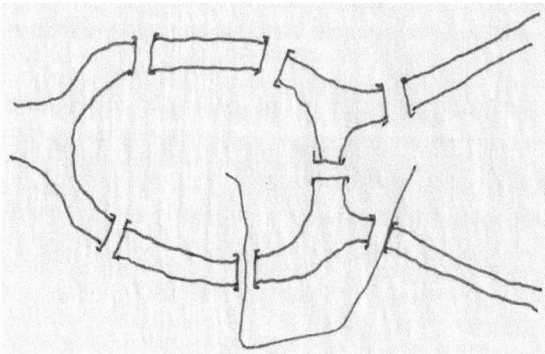

FIGURE 27   Going over island A to B (south) and then D (east).

Let's extend this – and go from *D* (*east*) to *C* (*north*). We would represent that journey as *ABDC*. Try the various journeys yourself, go in your mind from one land mass to the next. Now we can pause and think. You will see that every time you cross three bridges, you will need four letters to describe that journey. The four-letter journey (*ABDC*) covers three bridges; five-letter journey (*ABDCA*) would cover four bridges. By same logic, to cross or cover seven bridges, you will need eight letters of landmass to denote that journey.

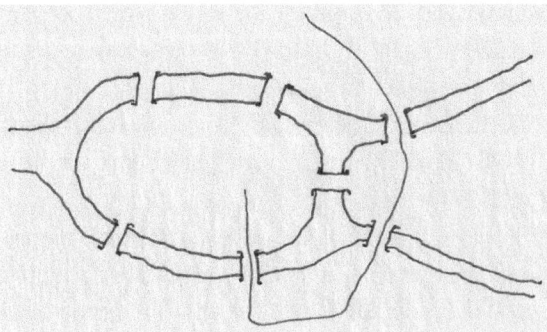

FIGURE 28   Doing the ABDC journey (island, south, east, north).

Let us go back to how many times letters *A*, *B*, *C*, and *D* could and would appear different number of times if one was to cross every bridge once and only once. *A* would appear three times (four bridges lead to it), *B*, *C*, and *D* twice each (each landmass has three bridges leading to and from it). But that means that one would need to step on our landmasses 9 rather than 8 times, which is what we agreed should be the number of times one should go over all landmasses, in total, if the seven-bridge journey could be undertaken!

That means that this particular journey could not be undertaken in the way we desired: to cross every bridge once and only once and cover them all in one journey. One could not cross the seven bridges of Königsberg without crossing one of them twice.

Conclusion is that the network of bridges (paths) and islands (or vertices as they would be called in mathematics) has to either:

- not contain odd vertices, vertices from which there are an odd number of paths

or

- contain two odd vertices, one at the beginning and one at the end of the journey.

For further meditation I recommend you plot your own favourite journey. As you look around next time you cross your favourite pathway, think of alternative little ways and whether you could cover all the favourite spots once and only once. I've been doing that on my bike rides. In case you'd rather stick to an imaginary pathway and doodle on a pad while you listen to your music, you can try the next exercise.

2.2 FURTHER DOODLING WITH PATHWAYS

Let's set some ground rules. The primary one is to exclusively use straight lines and points where the lines intersect. As noted previously, the lines we will call edges, the intersecting points vertices. We'll see what we can achieve by doodling with these simple elements. How complex can we go? The second rule is to never go twice over an edge, and always come back to the point from which we have started.

The one-point doodle is no fun.

FIGURE 29   One-point doodle.

The two-point/vertex doodle is a little more fun, but the edge that joins them doesn't come back to the first point from which you started without going over itself. So that's not great either.

FIGURE 30   Two-point/vertex doodle with a little path that joins them.

Ah, but three-vertex doodle is more fun. Now you get back to the first point, your vertex of departure becomes your vertex of arrival.

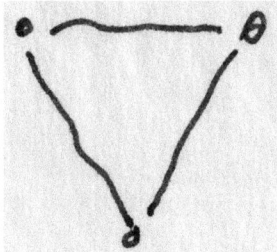

FIGURE 31   Three-vertex doodle.

And the four-vertex doodle will be even greater fun! Particularly if the fourth vertex is in the middle, allowing you to connect it to each other vertex but without any of the paths crossing. Don't worry, I won't go on like this to get to 20! And by 20 I didn't mean 20 × 19 × 18. . . . Seriously, what if the four vertices are given like this though? Could you connect them all making sure you never go over an edge more than once?

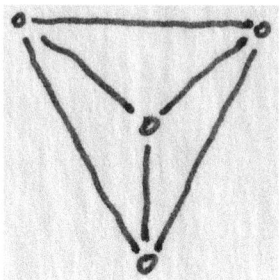

FIGURE 32   Four-vertex doodle with surprising topology.

You can imagine this particular doodle to be a projection of a three-dimensional shape, a tetrahedron. Now imagine that you are an ant (or some other appropriate little insect you like) for a little while. Instead of walking on the faces of this tetrahedron, you are walking only over its edges, trying to connect all the vertices as best you can, and covering each and every edge once and only once. But first, here's a tetrahedron, it has four faces, six edges, and four vertices.

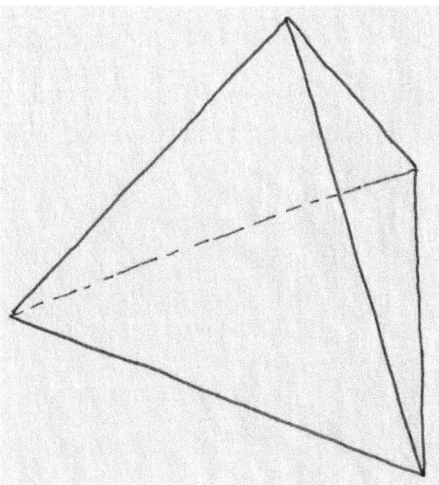

FIGURE 33   Tetrahedron.

The only satisfaction you gain from this exercise is to walk over all the edges once and only once, and to visit each vertex. But there is no limit how many times you can visit each vertex. You can add another restriction. Let's say we want to complete a cycle – to start from one vertex and to return to the same one at the end. In the next doodle, the path of going around the edges of a tetrahedron in one way is shown in red.

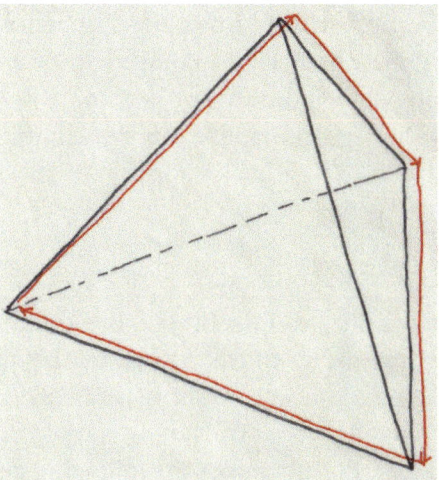

FIGURE 34   One possible pathway around the edges of a tetrahedron.

This wasn't greatly successful as we have two edges that haven't been covered. Try once more?

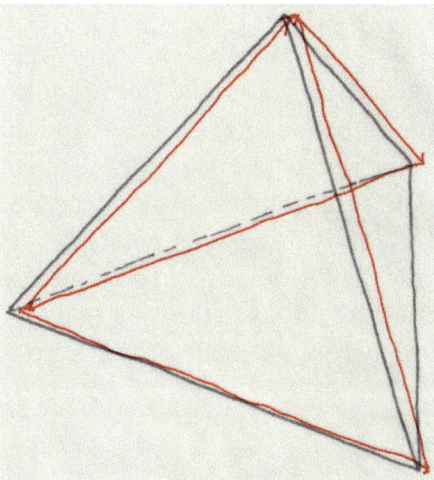

FIGURE 35   Another try of covering all the edges of a tetrahedron.

Will there be some conclusion to this? What happens with polygons or with polyhedra when we try to cover all their edges like this? A polyhedron is a three-dimensional shape like a tetrahedron – with flat faces that are polygons, edges, and vertices. If the polygons that make the polyhedron are regular, and only of one type, they make regular polyhedra.

On the scene comes Louis Poinsot. There's a bust in a little forgotten museum in the little sleepy town of Wisbech which was brought there by a wealthy merchant who managed to buy quite a few things that Napoleon Bonaparte had in his possession when he succumbed to illness and died at Longwood House, on the Island of Saint Helena in 1821. Poinsot's bust lives in a back room of Wisbech Museum, its old library.

2.3 THE INSPECTOR GENERAL

When Euler solved the puzzle of the bridges of Königsberg, the game with the vertices and edges didn't finish. Other mathematicians after Euler took it further. Among them was Poinsot (1777–1859). He was appointed Inspector General of the Imperial University Napoleon founded in 1808 (it was disbanded in 1896). Poinsot posed the following new problem in 1809, when he wrote this note:

> Given some points situated at random in space, it is required to arrange a single flexible thread uniting them two by two in all possible ways, so that finally the two ends of the thread join up, and so the total length is equal to the sum of all the mutual distances.

Poinsot came up with these exercises whilst the war was raging around him. Is doodling also keeping your mind steady and cool in those circumstances? Poinsot started by looking at a circle and marking on it points equally spaced. Then he would join them with segments. Then play a little more and introduce new rules as he went along. Let's have a look at a circle with seven and eight evenly distributed points. It's useful to distinguish between these because they are of odd and even numbers of points. Maybe there will be a difference of results to this type of doodle depending on the number of points the polygon is made of? Let's investigate a little more closely.

We can doodle the circle divided into seven equally distributed points in three ways.

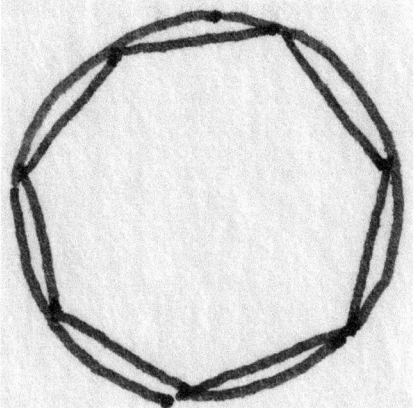

FIGURE 36   Heptagon doodle – connect all points in order to get back to start.

Then there is a heptagon doodle where you connect all the points in one uninterrupted stroke of a pen, skipping every second vertex.

FIGURE 37   Heptagon doodle where you connect all points but skip every second vertex.

And finally, a heptagon doodle where you connect all the points in one uninterrupted stroke of a pen, skipping two vertices at a time.

FIGURE 38 Heptagon doodle where you skip two points until you get back to the first vertex.

But if we have an even number of points, we can do an additional diagram of connections. So, like previously, let's do the first doodle with no skipping, the second by skipping one vertex each turn, and then two, and then three each.

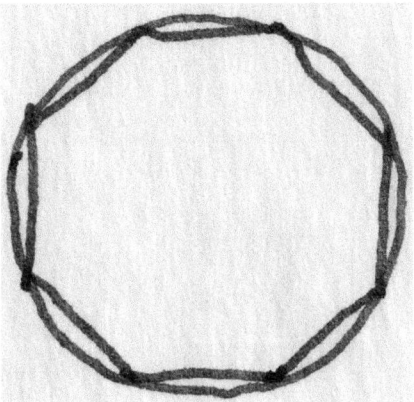

FIGURE 39 Octagon doodle with no skipped points.

FIGURE 40   Octagon doodle every second point is skipped.

FIGURE 41   Octagon doodle where two points are skipped every time.

FIGURE 42   Finally octagon doodle where three points are skipped at a time.

A couple of things happened while we were doing these doodles, based on Poinsot's ones. Firstly, with the seven points (heptagon) we can also do a little doodling with numbers. We can break number 7 in three ways:

$$4 + 3 = 7$$
$$5 + 2 = 7$$
$$6 + 1 = 7$$

Let's look at the heptagon, and as we start doodling we go through all points in turn as they come in.

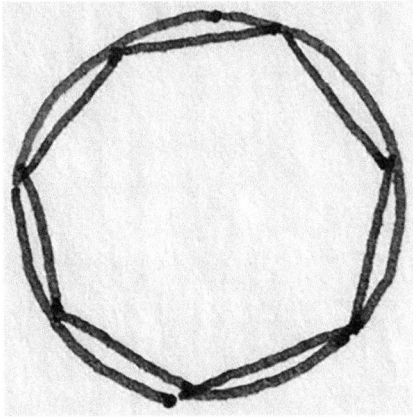

FIGURE 43   Heptagon doodle of least fun factor.

Soon we realise this is not too much fun, apart from seeing the heptagon itself. But let's then use the heptagon but skip every second point. Do this whilst not lifting the pencil off the paper.

FIGURE 44   Heptagon doodle with connecting every second point in one uninterrupted line.

And then the same but skip two points at a time.

FIGURE 45   Heptagon doodle in one uninterrupted line but two points skipped at a time.

What you would have noticed is that you start from some point and you will end back there, never going over the same path twice. In all three cases for the heptagon, we could do this: draw from one vertex, cover them all via our edges and get back to where we started.

We can check whether this works for even numbers of points too. Let's skip the ordinary octagon and go straight to the one where we first skip one point.

FIGURE 46   Octagon when we skip one point at a time.

Oops – we get a case where we can't go from the original point/vertex and cover all the edges and vertices in *one continuous* swoop. This type of figure is *degenerate*. But if we skip two points at a time we are back in the game – we can do it.

FIGURE 47   Octagon doodle in one continuous line with two points skipped at a time.

Now though, if we skip three points at a time we get another *degenerate* figure.

FIGURE 48   Octagon with three points skipped at a time.

Poinsot concluded that this type of doodle is possible to do only for an odd number of points, not for an even one. He generalised that when you have a polygon with n points (and sides) you will have N-number of new polygonal doodles where you can draw the polygon without lifting the pen off the paper, but that this number will depend on what our original polygon was like.

But this further also depended on the prime factors of a number of sides or vertices of a polygon.

The prime factors are those which are factors of a number, but are themselves prime numbers. Our two cases have different prime factors. Seven is itself a prime, and the number of new doodle polygons will be $N = \frac{1}{2}(n-1) = 3$. But when you have 8, which has only 2 as a prime factor, we can say that that prime factor is some $a = 2$, $N = \frac{n}{2}(1 - \frac{1}{a}) = 2$.

And in general with prime factors a, b, c, and so on, there is also a general formula

$$N = \frac{n}{2}\left(1 - \frac{1}{a}\right)\left(1 - \frac{1}{b}\right)\left(1 - \frac{1}{c}\right)\cdots$$

How useful is this then? It turns out incredibly useful as a property of polygons that have been used in network theory as well as statistical mechanics in our time.

2.4 CLOSING OF THE PATHWAYS WITH HAMILTON

William Rowan Hamilton (1780–1817) was an Irish mathematician, physicist, and astronomer. He came up with another alternative way of thinking about pathways. What if, instead of crossing every edge only once, you get through each *vertex* only once instead? He designed a game called *Icosian* (icosian comes from the number 20) which you can play by going over the edges of a dodecahedron.

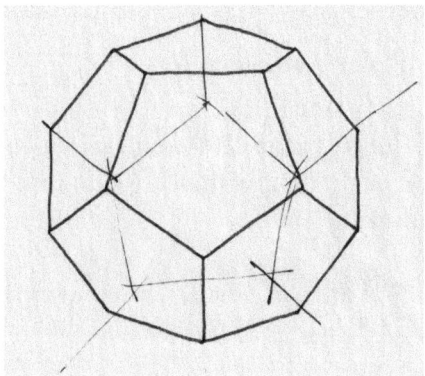

FIGURE 49   This is an image of a dodecahedron.

And if you flattened that dodecahedron, projecting it onto a two-dimensional diagram, you would get something like this:

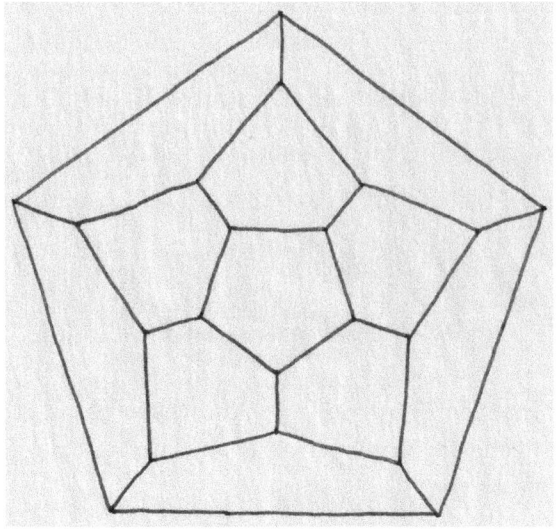

FIGURE 50   Icosian diagram game.

Icosian diagram is a two-dimensional diagram representing a dodecahedron as a map. I suggest you try this a few times. Apparently, it's a very easy game. Hamilton sold it to a game maker for £25 in 1857, at the time when you could buy a house for about £700 in Dublin or in London, so not bad. But the game turned out not to be a great success as it was too easy!

2.5 FURTHER EXPLORATION

Graph theory features in one of the most well-known films about mathematics, *Good Will Hunting* from 1997, staring Ben Affleck, Matt Damon, and Robin Williams. Although mathematics on that blackboard scene is not very difficult and certainly wouldn't count as genius material, it's still a good movie!

A further exploration on mathematics in movies may bring you to some other nice films.

One of the few papers that are scholarly but reasonably easy to understand and relate to the last part of this chapter is one by Branko Grünbaum, *Polygons: Meister was right and Poinsot was wrong but prevailed*, which is freely available from https://faculty.washington.edu/moishe/branko/ BG288%20Meister-Poinsot.pdf (accessed 8 June 2024).

CHAPTER 3

Diagrams to Contemplate Upon

S O FAR, ALL OUR meditations were based on some kind of walking or movement. But sometimes you will go to a place to just be still. Japanese Shinto temples (sometimes also Buddhist ones) between the 17th and 19th centuries often had what looked like a geometrical puzzle at their entrances. These diagrammatic puzzles on wood tables are called *sangaku*. They appeared on the temples around Japan during the Edo period. This was a period of Japanese near-total isolation from the Western world, and lasted from the very early 17th to the late 19th century.

The period of isolation was called *sakoku*; it was meant to regulate in a very strict way all relations with foreigners in order to restrict their influences on Japan. There were some exceptions to the rule: extensive trade with China through Nagasaki, Western technical and scientific innovations through what became known as *Dutch learning*, called *rangaku*, and diplomatic relations with Korea. But generally speaking, the country was isolated and a new type of knowledge was developed to replace mathematical knowledge that was developed in other places.

The sangaku is a result of *wasan* – a Japanese type of mathematics developed in this period. The learning circles were organised around a mathematician who was usually a samurai (hereditary military nobility). Sometimes learning circles would pose questions to others and exchange the problems – these were written on tablets and left in front of a temple or a shrine. Why this was so, we can't be certain. It seems it was done for

 DOI: 10.1201/9781003280668-3

the purpose of communicating mathematical problems but may also have been a kind of offering or an inspiration for a meditation.

So what can we learn from these geometrical diagrams? The diagrams were inscribed on a wooden plaque, with questions written around them. The problems were written in Kanbun, the form of Chinese used in Japan in this period and all the way until the mid-20th century for official and intellectual works. There are about, very roughly, 900 tablets that survive from this period. Some estimate that the total number of sangaku tablets made in this period was around 5000 – a nice number. But if we compare it with the number of people that lived in Edo (the future Tokyo) in the middle of the 18th century, about one million souls, then this doesn't seem to have been an incredibly popular thing to make, or rather it was an activity of the very few chosen, learned men. Some believe that the practice of suspending tablets in front of their temples can be compared to the European students in the Middle Ages posting their metaphysical thesis on the doors of their churches. In this way they would advertise their knowledge and achievement, as well as communicate to others what they wanted to discuss further.

The collection of tablets and the solutions to the problems posed on them sounded at some point like a good idea for a project. A mathematician Fujita Kagen (1765–1821) wrote a book *Simpeki Sampo* (*Mathematical Problems Suspended before the Temple*), and another one on the same topic, *Zoku Simpeki Sampo* in 1806. In these works he wrote about all the sangaku problems he came across, and explained how they could be solved.

We tend, perhaps, to mystify unknown images and practices, and our own culture's fascination with sangaku may fall into that category. There have been various attributions given to these tablets, their purpose and their meaning. One of these has been that they were made with a purpose of giving a focus for priests and learned temple-goers to meditate upon. What would such meditation bring? Very much what mathematical meditations usually bring to their practitioners. The contemplation on the beauty of geometry and of an order it reveals. Could this be the case? In so many years, and in so many places in which these tablets were displayed, there would have been people who contemplated upon them as they prayed in the temples in which they found themselves. So the original purpose did not matter much eventually – they *did* become tablets which inspired some people to meditate.

Let us then look at some of these sangaku problems.

3.1 THE RIGHT-ANGLED TRIANGLE WITH
AN INSCRIBED SQUARE

This is a problem from one of the sangaku from this period.

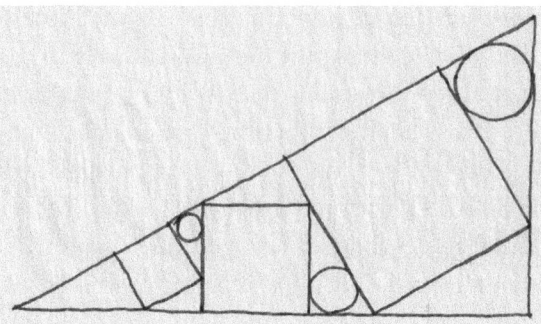

FIGURE 51   Sangaku problem – similar triangles and inscribed circles.

As in haiku poetry, there is not much there in terms of instruction. How do you think about it is up to you. You can of course see that the main triangle is a right-angled one. In it, there is a square inscribed. This square divides the main triangle into a larger part and two smaller parts. They are all triangles, and they are all similar to each other (and right-angled).

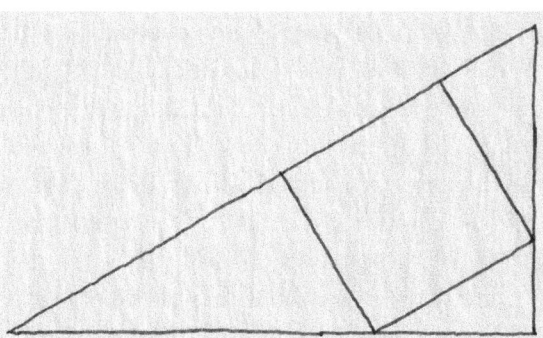

FIGURE 52   The main triangle with an inscribed square which divides the triangle into three further ones.

The three smaller triangles will all be similar to the original largest. The largest of these smaller, new triangles can also have a square inscribed in it. But then there are all these circles inscribed in the similar triangles too. Let us first decide what we will concentrate on. Later you may want to concentrate on something else and meditate on some other relationships in this diagram.

We get these three new triangles, with inscribed squares, which divide them into three further triangles, and in which, in the middle-sized of these three new triangles, there is an inscribed circle.

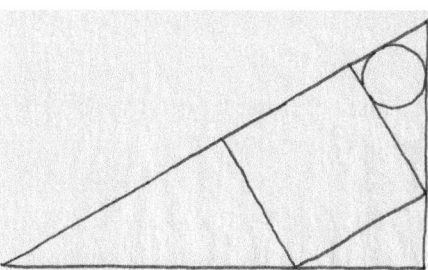

FIGURE 53   First, largest triangle, with inscribed circle in one corner.

Now, let us look at the second right-angled triangle, similar to the original one. It also has the circle inscribed in the similar manner to the original triangle. It just looks a little different as it is in a different position (orientation) to the first one.

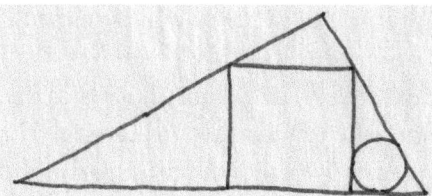

FIGURE 54   Second, smaller triangle with a circle inscribed in one corner in the same fashion as the previous.

Finally, the third right-angled triangle, similar to the original one, also has a circle inscribed in one corner.

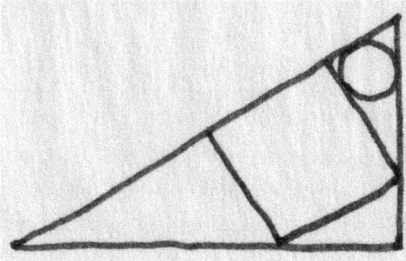

FIGURE 55   Third triangle with a corner circle in the same fashion as two previous ones.

All the triangles are similar to each other, as we already said. The two smaller triangles are nested in the largest one. That can be one meditation – how would you prove that they are similar? Remember that for two figures to be *similar*, they must have all the angles of the same size. Two similar figures, once they are proven to be similar, will have all the corresponding sides in a proportion too.

Why stop here with imagining and drawing smaller triangles inscribed in the ever larger number of triangles? We can go ad infinitum and imagine ever smaller triangles nested in the ones we create, with a little circle in one of its corners. If you are already having fun by thinking about ratios between various segments defined by these squares within these similar triangles, wait for the surprise questions.

And the main question in this: what is the relationship between the radii of these little circles?

Let us think through this. The ratios between hypotenuses of these new triangles will always remain the same, considering that they are constructed in the same way starting from the first triangle. So whatever the ratio is between the hypotenuse of the largest to the next triangle in our iteration, that same ratio will hold for a hypotenuse of the second to the third triangle, and so on. Iteration is a procedure that is repeated exactly in the same way it was originally done, but with the data you acquired through the last time you applied that procedure. That would mean that the ratio of corresponding sides of the triangles in different sequences of triangles would stand in a same ratio to each other.

Now let's look at it a little further still. We have a triangle sequence 1, 2, and 3. Each of these triangles has sides. Let z always be the hypotenuse in a series of triangles. The corresponding sides of all these triangles will be in the same ratio to those in the corresponding sequence.

This means that $z_1 : z_2 = z_2 : z_3$. And this in turn means that the hypotenuse z_2 is the *mean proportional* between z_1 and z_3.

3.2 THE MEAN PROPORTIONAL

But what is the *mean proportional*? A *mean* between two or more segments can be different things. There are several *means* in other words. The one that would probably come to mind first is the *arithmetic mean*. Arithmetic mean is the sum of two magnitudes divided by two. If there are more than two magnitudes, you would sum them all up and divide by the number of all magnitudes. We could draw an arithmetic mean by using a construction of a semi-circle. Add a and b and the semi-circle overarching them would have a diameter of some length c. This length is the arithmetic mean of the two values a and b.

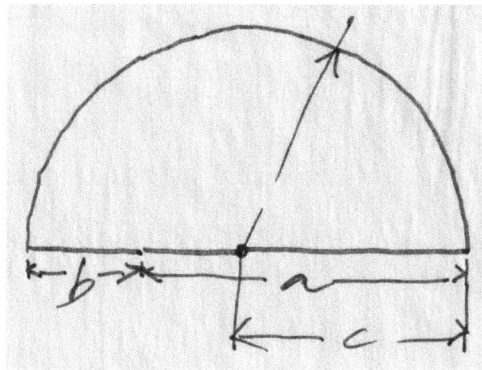

FIGURE 56   The arithmetic mean of segments a and b is the segment c.

That is, arithmetic mean would be some value c which could be calculated as $\dfrac{a+b}{2}=c$.

If there is an arithmetic mean, there must be a geometric one too. A geometric mean is a square root of the product of two magnitudes, $\sqrt{a \cdot b}=c$. We can represent that with a geometrical construction too. The product of two magnitudes is an area between them. So, we could set out two magnitudes as sides of a rectangle with sides a and b. That rectangle will have the same area as some square. How can you find the side of a square whose area is the same as that of a rectangle whose sides are a and b? Finding such a square or finding in effect a square root of $a \times b$ has been known to ancient civilisations in various forms, including how to calculate the square root of two. Mesopotamians knew it, for example.

Let us look then at how we could show a geometric mean using a triangle whose hypotenuse is the sum of a and b.

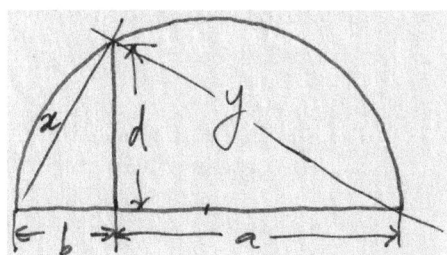

FIGURE 57   Find the geometric mean.

The only thing we need to know is the so-called Pythagoras' Theorem: the sum of squares on two smaller sides of a right-angled triangle equals the square on its hypotenuse. Looking at our diagram we can say that

$$(a+b)^2 = x^2 + y^2 \tag{1}$$

But both x and y can further be expressed as sides of two smaller right-angled triangles,

$$b^2 + d^2 = x^2 \tag{2}$$

and

$$d^2 + a^2 = y^2 \tag{3}$$

If we connect the first statement (1) with (2) and (3), we get

$$(a+b)^2 = b^2 + d^2 + d^2 + a^2$$

Now, we can multiply the bracket out from this statement which would give us

$$a^2 + 2ab + b^2 = a^2 + d^2 + d^2 + b^2$$

With some tidying up, that would come to $d^2 = ab$, which means that $d = \sqrt{ab}$. (We won't worry too much about possible negative values here.)

Let's now go back to our sangaku triangle. It turns out that all the lengths (but not areas!) in these triangles are related in a similar fashion. The radii of the three circles are in fact related to each other in the geometric mean ratio.

3.3 NOW FOR THE CIRCLES

Now, if you were to draw a little triangle within each of the second (in size) triangle,

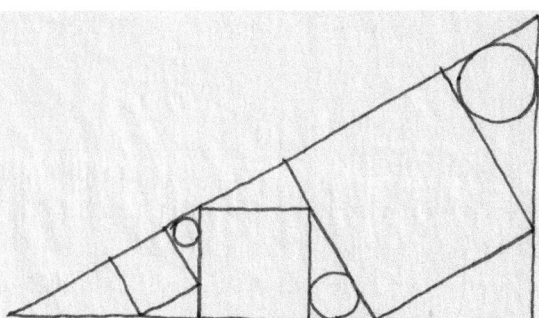

FIGURE 58   Back to our main sangaku diagram.

you would see that they are also in some kind of ratio to each other. Because the hypotenuses of the triangles are in this specific ratio of mean proportional, any lengths within them positioned in a way that is consistent with this will also be in that ratio. This means that the radii of the circles constructed in our sequence of triangles are also in the mean proportional with each other. In other words, we have three circles for which their radii satisfy the equation

$$r_1 : r_2 = r_2 : r_3$$

The *mean proportional* is therefore the middle value in a ratio of three values. Its magnitude is also a square root of the rectangle that has the sides of the same size as extreme values in that proportional.

3.4 SIGNS BY THE ROADSIDE ON A SHINTO PILGRIMAGE

A famous book *Signs by the Roadside* is a collection of notes written by Ivo Andrić, an author from my original homeland. He won the Nobel Prize for his novels (in 1961). This little book is a summary of his daily meditations. Would he have liked for them to be published one day as a public record? After all, these were his daily thoughts, his signs on the road of his search for meaning. But what if, instead of sharing our own intimate short meditations, we follow a universal one? If we look at the signs someone leaves by a *roadside* for us to meditate upon would we all have similar meditations? Even if sangaku tablets were not necessarily meant to fulfil this purpose, we can well imagine that there was some universal role here in understanding the universal mathematical truths portrayed in their diagrams.

3.5 LITTLE MEDITATIVE QUESTION FROM DYSON

Freeman Dyson (1923–2020) was interested in sangaku. He wrote a foreword to a book on sangaku noted in 'Further Reading'. There he said that everyone should have in their mathematical library some of these problems. There are people out there still making such problems – perhaps one such problem is so enchanting it would be worthy of being made into a new tablet to adorn your room.

Another thought to leave you with is the question of the universality of mathematics. Is mathematics universal? The laws of it are – but its manifestations certainly have a huge diversity over different cultures, places, and historical periods. Confucius, who influenced Shintoism and Buddhism said, "You should devote all your time to study, forgetting to have meals and going without sleep". On a sangaku made by Saito Kuninori, positioned in Kitamuki Kannon Temple near Ueda in Japan, and hung there in 1828, this suggestion by Confucius is continued with the following inscription:

> His words are precious to us. Since I was a boy, I have been studying mathematics and read many books on mathematics. When I had any questions, I visited and asked mathematician Ono Eijyu. I appreciate my master's teachings. For his kindness, I will hang a sangaku in this temple.

It may be that the new sangaku you make is to do with some mathematics you learnt from your best teacher.

3.6 FURTHER READING

Sacred Mathematics: Japanese Temple Geometry, by Fukagawa Hidetoshi and Tony Rothman (and with foreword by Freeman Dyson), was published by Princeton University Press in 2008.

I would highly recommend the novel by James Clavell, *Shōgun*, published first in 1975. There were two TV series made after it, the latest one in 2023. That also comes highly recommended. The story is about the Edo period and power relations. I have checked neither the novel nor the series for their historical accuracy, and the recommendation comes only because of their unadulterated cultural entertainment value.

CHAPTER 4

Music for Your Mind

THE MEDITATIONS IN THIS chapter are based on recognising the relationships between mathematics and music. The beauty of music is that you can enjoy it without knowing how to recreate it. Very few people have amazing voices, yet even those of us who don't, can experience some joy by just listening to music others make. We can also enjoy music by working out some mathematical structures within it. And there is plenty of mathematics in music.

Perhaps there are some easy tunes we could study, but we will start with Bach as the structure of his pieces is very mathematical. As everyone who knows anything about music will tell you, he is the teacher of many musicians who came after him. By doing a little bit of this analysis, let us also explore the possibility of contemplating music by using mathematical relationships. Music of your mind, in other words.

4.1 BACH, THE ART OF FUGUE, AND GROUP THEORY

Johann Sebastian Bach (1685–1750) was an organist and a composer. To us he is famous as a composer, as humanity didn't have recording devices in his time. But Bach being so well known was a consequence of his work being rediscovered by Felix Mendelssohn (1809–1847) in the 1820s.

Bach's music is one of the polyphony. We can enjoy this without being able to understand anything of the structure underlying it, but it may bring you a little more joy if you do understand some of its principles when you next hear it.

Polyphony is the interplay of multiple voices. You can start with two voices, but go to as many voices as you like (and as many as would fit in

DOI: 10.1201/9781003280668-4

nicely with your tune). The only thing that matters is to create a combination within a melody of a coherent sound. Remember "Frère Jacques (dormez-vous)"? This is a little song that many children will revel in learning in their French lessons around the world, including little French children. Bach's music followed this type of method of constructing a melody.

Canon is a single theme played against itself (relates to our French children's rhyme). Copies of the theme will be staggered in time but also sometimes in pitch: you can have a main theme sung in say C major, and the next in G major. Fugue is similar to canon, but is more complex in structure and less rigid in following the rules. You introduce many different further elements. You can introduce a different secondary theme. When all your voices are joined, there are no rules, and composers can do what they like in a fugue. There is no formula.

Composers can deviate from the main principles of the simple form of polyphony, and the voices can be slightly different in both sounds and rhythm. But if you try doing this yourself, you will soon discover that the exercise of creating such a melody is highly complex. And if you continue with developing complexity of your piece, you will get an idea that you can use the first voice as a theme. This can then be repeated in different ways – in different keys and rhythms.

Bach's probably most favourite form of music was a fugue. One of his famous pieces is *The Well-Tempered Clavier*. I used to put my daughter to sleep when she was a child by playing it to her. I am not a great pianist, but it is a relatively easy piece, and very soothing. This piece consists of 48 fugues in total. Bach's last, unfinished piece, was called *The Art of Fugue*.

Now this further exercise is for the lovers of mathematics and the lovers of music. It is by no means intended for those who are experts in the field only. It is here just to try and show you how beautiful structures that are shared by mathematics and music can be studied through modern mathematics.

4.2 OUR TOOL SET – GROUP THEORY

Just to remind you so you don't forget in the meantime, we will now look at a possible way of creating a structure that will allow us to play with sets of voices in some way. There will be plenty of opportunities to ask yourself – why is this here? It is so that we can get a sense of a canon and of a fugue in their mathematical interpretations.

The first thing to remember is the concept of a binary operation. You want to add or multiply or subtract two numbers, or do anything of a sort where some kind of operation is imposed on two things of the same type, like numbers. (Later you can replace numbers with notes or even with a collection of notes, like a simple theme.) The operations are in themselves a wonder of mathematics. They have been around since prehistoric humans made some incisions on baboon bones like those found near Ishango Lake in Congo to show how to add or double simple numbers. What has evolved over the millennia is the more modern way of looking at operations and the numbers you can apply these operations over. The operations don't have to be arithmetical operations any longer, and the set of the same types of things over which you apply them don't have to be numbers. Here we enter the world of group theory.

For a group to exist, you have to have a set of elements, like for example numbers. Let's stay with numbers for a minute as they are easiest to work with. Let us say our numbers are whole numbers only, and we want to limit ourselves to adding them. So we can say that this set of numbers and this operation form a *group*. Our group consists of whole numbers and a binary operation of addition.

There are many different types of groups depending on the choice of both: the set of elements, and the binary operation you chose to implement on them. For our purposes we want to get to the point where we can think of a group that will have a set of elements of themes or tunes, and an operation that we can use to combine them. We are not there yet. Let's review first some characteristics of groups.

There are certain rules you have to follow if you want to have a group, mathematically speaking.

Rule #1: You have to have an element, called *identity*, that will act as a neutral element. Let's say we are using a set of numbers, and our operation is addition – the *identity element* is 0 because if you add 0 to any other element from your set, you don't change that element. The identity element, when used with the chosen operation, does not make a change on any element from our set.

Rule #2: You have to have an *inverse* element that corresponds to each element from your set, and that will give you your identity. In our case of whole numbers, you can add 1 to −1 to get 0 (our identity element). So the inverse element for each of our elements will be the negative of the original, and this is valid for every whole number from our set of whole numbers.

Rule #3: Your binary operation has to be *associative*. That means that elements can be associated in any order: $(a + b) + c = a + (b + c)$. That's pretty straightforward and is a characteristic of our binary operation.

Rule #4: Your binary operation must be *commutative*. This means that if you chose any two numbers from our set of whole numbers, the way you add them with our binary operation doesn't matter. In other words, or rather, digits, $1 + 4 = 4 + 1$ and so on for any two numbers you choose from our set.

So now we have a concept of a group sorted out. Let us see how this works with music, and more precisely, Bach's fugues.

4.3 APPLYING GROUP THEORY TO MUSIC

Group theory can be applied as most mathematical constructs can, not only to working further in numbers but to geometrical objects. In this vein, let us apply group theory to the study of symmetry, and that of a symmetry to the one we can find in music.

Let's see how we can visualise symmetry of a group that constitutes Bach's fugue, in particular his *The Art of Fugue*, his last and unfinished piece of work. We will only look at the technique of *transposition* and *inversion*.

Transposition is when you move your motif around. For example, if you play a piano, you can play a motif in one key, then you can transpose it to another register (higher or lower). On the other hand, the inversion is like a reflection of a melody or a motif about a fixed axis. We are looking at piano, although *The Art of Fugue* was written for harpsichord, which is an ancestor of a piano. It is also played by keys, but since there are very few harpsichords in existence today, *The Art of Fugue* is mostly played on a piano.

The next task is to connect the music to groups. First, we need to look at an octave. It is divided into 12 pitch classes. They are (if you look at the piano) notes we call as follows:

FIGURE 59 Piano keys.

But we can also create a more abstract way of presenting them. For example, we can create a table of values and assign to them our notes. You will see that A# is equal to Bb.

FIGURE 60   Piano keys in a table.

Now we will get on with the further work of generalising this further. We will replace these keys with numbers. Let us assign number value to all these notes. Let us say we give C value of 0, C# of 1, and so on. In this way we bring music a little closer to mathematics. And if we say that transposition and inversion are some kinds of symmetries, we can get even further to our mathematical structure of a group.

Now this is also getting close to geometry – symmetry may be easier to explain with geometry. To see how *this* works, let's introduce the symmetries of a square.

Square has two symmetries – four rotational and four reflective. So let us create a diagram showing the eight rotational and reflective symmetries of a square.

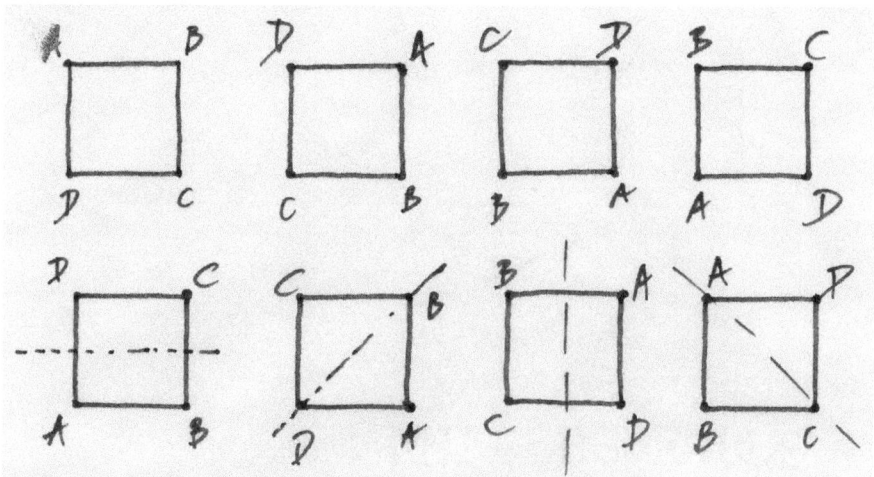

FIGURE 61   Eight symmetries of a square.

You can keep the square as it is, the first top left-hand-side figure (with a slightly smudged *A*). That will be the identity element of our (dihedral) group of symmetries of a square. The dihedral stands for these two symmetries. Then you can turn it around (using rotational symmetry) and reflect it in its axes of reflection. Altogether, in our group, there are eight possible elements and one operation – the operation of combining two symmetries (reflection and rotation). We can create a table for all the possible outcomes of combining these symmetries. Such a table is called a Cayley table, after Arthur Cayley (1821–1895) who first mentioned such a thing in his 1854 paper on the theory of groups. But constructing such a table is a little complicated for an exercise in meditation that is not actually about tables. So here is how a Cayley table for the dihedral group of four for square looks like, just so you know. The elements are the eight squares we have in the previous image, and their symmetries are as follows:

- e is the identity.

- rotations r, r^2, r^3 are of 90°, 180°, 270° around the centre of the square, and going in an anticlockwise direction.

- the reflections t_x and t_y are reflections about the x and y axis respectively.

- The reflection t_{AB} is about the diagonal through vertices A and C.

- The reflection t_{BD} is about the diagonal through vertices B and D.

And voila, we get a Cayley table filled with all the combinations of rotational and reflective symmetries of a square.

	e	r	r^2	r^3	t_x	t_y	t_{AC}	t_{BD}
e	e	r	r^2	r^3	t_x	t_y	t_{AC}	t_{BD}
r	r	r^2	r^3	e	t_{AC}	t_{BD}	t_y	t_x
r^2	r^2	r^3	e	r	t_y	t_x	t_{BD}	t_{AC}
r^3	r^3	e	r	r^2	t_{BD}	t_{AC}	t_x	t_y
t_x	t_x	t_{BD}	t_y	t_{AC}	e	r^2	r^3	r
t_y	t_y	t_{AC}	t_x	t_{BD}	r^2	e	r	r^3
t_{AC}	t_{AC}	t_x	t_{BD}	t_y	r	r^3	e	r^2
t_{BD}	t_{BD}	t_y	t_{AC}	t_x	r^3	r	r^2	e

FIGURE 62   Cayley table of the reflections of square.

And that is enough for the moment to start you thinking. It is obvious from the table to see that there will be 64 outcomes if one combines these two reflections. Every time you combine two symmetries, a resulting square will always be one of the squares from our picture. These cover all the possible symmetries of a square.

Now let us turn these principles to see how you can use such type of thinking to create music. With the square you already got used to turning things around (rotation) and reflecting things in axes of symmetry. Let's now look at the 12 notes and create some kind of visual structure that we can play with.

4.4 THE MUSICAL CLOCK

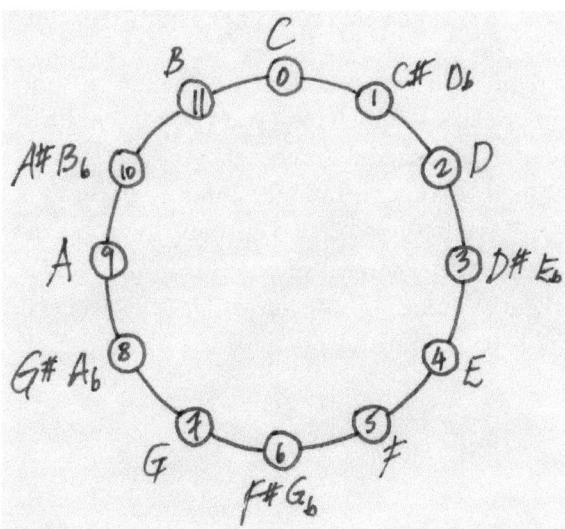

FIGURE 63   Musical clock, with notes from one octave positioned in a circle.

This little device we will call the musical clock. You could turn it around, clockwise and anticlockwise, and you could also reflect it in its axes of symmetry. Where those axes will be is up to you. You can have horizontal or vertical, or you can note the opposite notes and use that line as an axis of symmetry.

If you turn the musical clock around, what you are really doing is *transposition*. Let's say you have a little melody, a motif. Now turn the *clock* around and start it from another tone. Bach used this in his fugues: a transposition by turning a musical clock by a whole number +mod12. The "mod" stands for modulo – meaning that for each tone of the original motif, there would be a corresponding one +12. But you can use it for turnings of other modulo – for example 6. Then a note C would become F# or G♭. It is up to you. You can play. Create a little tune and transpose it by turning around the musical clock.

How many times can you do this? Infinitely many times. You can create a song to last an eternity. Every time your original motif, your first melody, would be repeated in a different pitch.

Now let us connect what we have done here, with the concept of *inversion*. We mentioned it earlier – it is when you do a reflection of some kind. We'll do this too on our dodecagonal musical clock. The next motif can be a collection of three notes: say C, E, G.

(If you don't have a piano, and your singing is not up to scratch, you can always use one of the online piano apps and get it to play in different voices.)

To use inversion, you can reflect a line of notes in one of the axes of symmetry of your clock – which axis you choose will be up to your taste. Axes of the regular dodecagon connect its vertices, so it could be the 0–6 axis. It is imagined – you can meditate on any other axis to use for the same purposes. This will give you two options that you can use in different ways – C-E-G, C-A-F, C-F-A and so on. This musical clock is the same as the one from previous diagram, but with a reflective symmetry with an axis between note 0 and note 6.

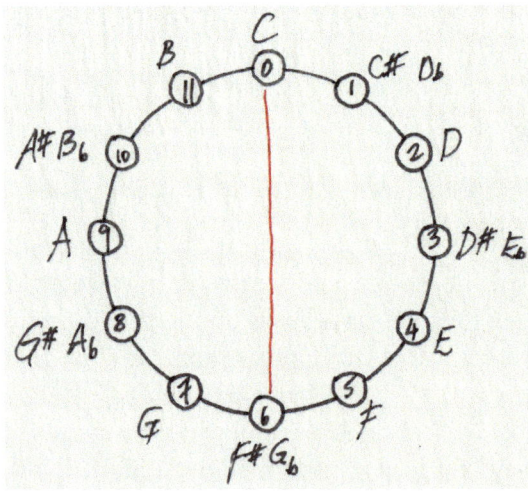

FIGURE 64   Musical clock with axis of symmetry 0–6.

It transpires (no pun intended on the transposition) that these 12 notes, and the symmetry operations we employed, also constitute a group in a mathematical sense. Like with the square, we could use a combination of transposition and inversion to get different tunes when composing our little fugue from an original melody.

The thing to let yourself wonder a bit about perhaps is this: how does it happen that the beautiful music of Bach's fugues so well fits the mathematical concept discovered a couple of centuries after he created this music? Group theory began to be developed in the 19th century, but it was the German geometer Felix Klein (1849–1925) who used geometry and symmetry groups to develop a method of characterising different geometries based on group theory.

Mathematics and music have this kind of common feature – when we recognise their laws, we experience beauty. Their structures sometimes coincide. The recognition of beauty and the enjoyment comes also through recognising these structures and their coincidences.

Now you will perhaps be able to make some new music, or just enjoy letting your mind go around that musical-mathematical clock. Next time you hear Bach's fugue, you will be able to hopefully recognise some of the transpositions and inversions and even possibly visualise them through symmetries we discussed.

4.5 FURTHER EXPLORATION

There's a nice paper in *Scientific American* titled 'Secret Mathematical Patterns Revealed in Bach's Music', which you may be able to access. Link is here (last accessed 14 June 2024): www.scientificamerican.com/article/secret-mathematical-patterns-revealed-in-bachs-music/

There is a classic book many mathematicians and mathematics teachers have read and loved, so you may want to try it too: *Gödel, Escher, Bach: An Eternal Golden Braid* by Douglas Hofstadter. It was published in 1979 by Basic Books, NY.

CHAPTER 5

Friendships

Mathematicians and Musicians

THIS MEDITATION IS ABOUT musical ratios and numbers, and how they were considered in the 14th century by a musician and his mathematician friends. There's a lot of numbers here, so for those who are not number lovers, skip to the next section in this same chapter – it features a little more of the thoughts one of these mathematicians formulated.

We start with Pythagoras or his followers. They noticed a long time ago that three intervals produced the most pleasing sounds when a chord was divided and then struck to produce a sound: an octave, a fifth, and a fourth. They can be represented as ratios of lengths 2 to 1, 3 to 2, and 4 to 3.

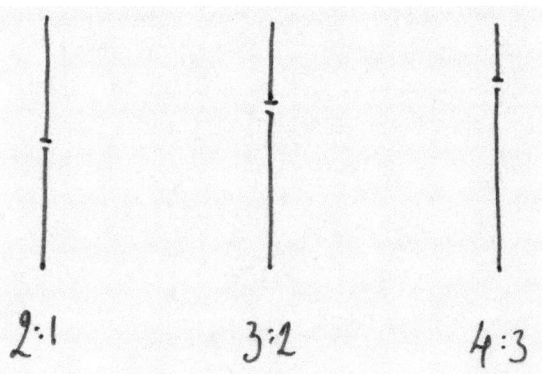

FIGURE 65   Music ratios of harmonic pairs.

   DOI: 10.1201/9781003280668-5

These are therefore three sets of two numbers in a ratio such that they differ by one only.

Then in the 14th century along came a musician who noticed this again. There was mathematics there, and he wondered whether there were any further such ratios in music. He wasn't sure what was going on. This was Philippe de Vitry (1291–1361), a musician and theologian. De Vitry became a bishop of Meaux near Paris in 1351, and stayed there until the end of his life. He was well liked at the papal court at Avignon and knew Nicole Oresme (1320–1382), a famous mathematician. Oresme was a bishop of Lisieux and also studied at the University of Paris.

So the two knew each other, and must have spoken about music. Oresme wrote a book on mathematical proportions, *Algorismus proportiounum* (after 1351) and dedicated it to de Vitry. In this book Oresme likened de Vitry to Pythagoras – both were indeed interested in ratios and in music. Yet de Vitry asked another mathematician, then an unknown scholar, and a rabbi from the south of France for further help. He wanted so see whether there were some other ratios like those he found in the octave, a fifth, and a fourth. He was interested in these pairs of harmonic numbers as he developed a particular liking to their sounds and wanted to find whether there are any more such pairs so that he could use them in his compositions.

Looking at the structure of these *harmonic* numbers, he added the fourth, so now there were the four pairs, 1 and 2, 2 and 3, 3 and 4, and 8 and 9. They can all be described as products of their prime factors 2 and 3, $2^n \times 3^m$.

The table of such numbers up to 1000 gives you some ideas for smaller such numbers. From here you can see that we can't find more than four such cases that were already known.

De Vitry asked now a rabbi and mathematician, Levi ben Gerson, known also as Gersonides (1288–1344) to help him. Gersonides was a theologian and philosopher. He wrote exclusively in Hebrew. The only one of his works to be translated in his lifetime into Latin was this treatise he wrote for de Vitry, on harmonic numbers.

Gersonides found that these four pairs of numbers were indeed the only cases of harmonic pairs there are. How did he do that? Gersonides conjectured this: if he could prove that you don't have a pair that is larger than 8 and 9, you have proven that indeed these known pairs are all there is to it.

TABLE 5.1   Table of Harmonic Numbers in Music, up to 1000, as described by Gersonides and de Vitry.

Multiplied			Powers of 2 ♫				
Powers of 3 ♭	1 (2⁰ also 3⁰)	3	9	27	81	243	729
	2 (2¹)	6	18	54	162	486	
	4 (2²)	12	36	108	324	972	
	8	24	72	216	648		
	16	48	144	432			
	32	962	288	864			
	64	192	576				
	128	384					
	256	768					
	512						

He conjectured like this. A harmonic number is of the form, as we said, $2^n \times 3^m$. In harmonic pairs, the difference is one. If there are two numbers of this form, and they are different by one, then one must be even and the other odd. We are going to show something, like Gersonides did, eventually. But first let's look at some numbers individually.

Now look at the table and you will see that the only odd numbers in the table are the numbers that are powers of 3.

Why is this?

You can stop reading here and think a little.

But in case you want to continue, keep reading. Let's say a number n can take any form, and the next number will be $n + 1$. So one of them is even, one odd. Any even number can be written as say $2a$. The product of a number from a harmonic pair is where one is even, the other odd, so that would look like this:

$$n(n+1)$$

Now we can replace this with a new formula, as we said our n is even, so we will write it as $2a$, for the even number. The product of an even and odd number can then be written as:

$$2a(2a+1)$$

You can now multiply this out and you would get

$$4a^2 + 2a$$

Now we can take 2 as a common factor. This means that our product will always be even, as every even number is divisible by 2:

$$2(a^2 + a)$$

Now the doubting amongst my readers may say, "but what if the first number was odd?" That means that $n+1$ is even. Let's try that. Start from our original product of two numbers,

$$n(n+1)$$

But now we say that our $n+1$ is even, so the n will be an odd number. In that case, if you repeat the prior formula but replace it for $2b$ rather than $2a$ we get

$$(2b-1)2b$$

And when you multiply that out and then take the 2 in front of the bracket, you get

$$2(b^2 - b)$$

Again, if a number can be divided by 2 (it is a factor of 2) then that is an even number.

We have just proved that the product of an even and odd number is always even.

This was something Gersonides would have known. And if we now go back to our table and see that the only odd numbers will be those which are powers of 3 multiplied with 2^0, that means that we can form two equations. Let's see how that works.

It means that one of the two harmonic numbers must be a power of 3 (multiplied by 1 as a 0th power of 2) and the other a power of 2. If you are confused here you won't be the only one – this is what bothered de Vitry too. How so? Well, the original definition of our harmonic numbers is that they are always a product of powers of 2 and 3.

Let's get on and hopefully it will become clearer.

We can form two equations now. One is when 2^n is larger by 1 from 3^n. Remember that this is because 3^n must be one of the numbers in our harmonic pair. One possibility would be that

$$2^n = 3^m + 1$$

And the other possibility is that 2^n is smaller by 1 from 3^n, which means

$$2^n = 3^m - 1$$

There are eight whole tones between C and the next C. You can work out the remainders after division of powers of 3 by 8, and powers of 2 by 8. Taken the example that $27/8 = 3$ remainder 3. For powers of 3, in fact, the remainders all turn out to be 1 or 3, depending on whether the power is even or odd.

Now look at dividing powers of 2 by 8. The reminders are 1, 2, 4, then 0 for all powers higher than 2.

For those who know about modulo arithmetic, this is nothing more than working out in modulo 8. It's as if you don't have numbers 9 and 10, but only numbers up to 8 and then you start again.

Let's then look at all our possibilities.

For $2^n = 3^m + 1$ we look for values when m is odd. When that is the case, $2^n = 3^m + 1$ must have a remainder 4. So, $n = 2$, and $m = 1$. This will give us a pair of 3 and 4.

When m is odd, the equation will give us the consecutive numbers 1 and 2.

For the other equation, $2^n = 3^m - 1$, we look first at m being odd. When this is the case, 3^m has remainder 3, and therefore the $2^n = 3^m - 1$ would have to have a reminder 2. As a result $n = 1$ and $m = 1$, so that will give us the pair 2 and 3.

Finally, for this same equation $2^n = 3^m - 1$, we look when m is even. If this is the case, we can apply our old trick and say that m can be equal to some $2k$. So we now have

$$2^n = 3^{2k} - 1$$

Whenever you have a difference of squares, it is of the form $3^{2k} - 1 = (3^k - 1)(3^k + 1)$. That gives consecutive harmonic numbers of 8 and 9.

There are no other options!

And with these numbers we finish our meditation on mathematics and music and their relationship with numbers. But the chapter is not over yet – see a little more about our friend Gersonides in what follows.

5.1 GERSONIDES AND THE FREE WILL QUESTION

This section is mainly for those interested in logic, although it can be interesting to those interested in philosophy and theology apart from mathematics. I hope you enjoy reading this and meditating upon it.

Gersonides worked on mathematics, logic, and astronomy mainly until he was 37 years old. Not unlike some other mathematicians (Blaise Pascal and Maria Gaetana Agnesi, for example), afterwards he wanted to examine more closely the theological questions. One of those was focused on the question of free will.

If you are an atheist this may put you off, but bear with the story just for a moment. Or you may be an atheist who is an atheist because they can't believe that a divine entity would allow horrendous crimes to be committed and hence there is no God (most of those professing to be atheist, in my experience, quote this reason). In either case, there is much mathematics here that will surprise you.

There are two basic opposing views in the debate about free will. One is that if there was free will, it would not really be compatible with a deity whose own will would *trump* (with a small t) all other wills in the world. The other view is that there is no free will, so those who commit crimes are in a way powerless as everything is predetermined (by this overreaching deity, that some call God). Those are extreme opposites, and as in any debate, there is a myriad of different views in between these two.

Gersonides offered an interesting and a very novel view on this question. His view was also interesting in terms of teaching. He believed that love of learning was crucial for good life and even immortality. Briefly, his argument went as follows. The acquired intellect (not the one that is part of the divine) corresponds with the divine intellect. Because of that, it is incorruptible. True knowledge is therefore incorruptible. And knowing mathematics would help you on making your mind about free will.

Consequentially, the person who is able to achieve true knowledge is able to achieve divine-like immortality through learning.

But what about that first thing we mentioned? The free will?

A common question among many philosophers of his times was an argument about the omniscient deity (who knows everything). If such deity knew everything, then they would know everything independent of time as well as place. They will know whether something has happened, is happening now, or will happen in the future. In that case, what place is there for an individual to do something that is not to such a deity's liking?

If everything is already known, what can an individual human do apart from following their predetermined destiny? Gersonides solves this through the comparison with mathematics. His argument goes something like this.

If a deity knows all events, they would know them before or during the event (as well as after). If the deity is all powerful, all these events will be foreknown and can be changed. The knowledge of such divinity is therefore about knowledge of everything that will ever happen, because if there was no such knowledge, then divinity is not really all-seeing and all-powerful. And if something happens that such a deity hasn't really considered, then they are not all powerful. So far so good.

But what if there is knowledge of general and knowledge of particular things? And what if these are very different in nature? And if our deity accepts this *because* it is about some type of knowledge that is mathematical and that determines the nature of things as they *actually* are? The divinity (or predetermination) can be about all events that can in general happen, but there is still something that is unpredictable.

Mathematics comes to help us have some small, but important aspect of a free will. To prove this – that we have a little bit of free will – Gersonides compared knowledge of a deity with the understanding and knowing of a continuous magnitude. Continuous magnitude is that which can be indefinitely divided. If our divinity knew of *every* possible particular, then they would know of all the divisions of such continuous magnitude. But these are *infinite*. Surely divinity would not want to go against their own laws of nature and mathematics?

5.2 FURTHER EXPLORATION

Highly recommended as a follow-up of this chapter is to listen to any music by Philippe de Vitry of which there is much already on the internet, but you can also find some on music platforms.

I recommend anything you can find on Gersonides. There is a lovely novel by Iain Pears, *Dream of Scipio*, in which all of this is explained in the most wonderful way. You need to persist with it a little, as Pears takes his time to bring about the story. But you will be rewarded in a multitude of ways.

CHAPTER 6

Presence and Transience

A ND SO FROM THE previous, we come to the current chapter. From Gersonides and his appearance in Iain Pears' *Dream of Scipio*, let's explore some other themes from literature that contain (or we can find references to) some mathematics.

There have been many papers and now a few books written on the links between mathematics and literature. Such works usually cover two major cases: firstly, you can use mathematical structures to underpin your story, and secondly, you may use some of the motifs from mathematics to put them in your story. This may include writing about people who do mathematics.

What we will do here is a little different. Let's begin to sketch this meditative journey.

I will start with my favourite writer, Aldous Huxley. It is strange how one chooses people and things one likes – there is much of rationality but much more irrationality about it. My favourite writer wrote a book about one of my favourite colours. He was such an honest philosopher and writer that he is closely approaching, as if by some invisible asymptote, philosophers who had a more mathematical bend in their opus. Huxley was a master of the English language, writing beautiful and provocative novels that explored themes of what the world would look like if it had taken the turn to materialise our ideas of the best possible worlds into reality.

One of his pupils was George Orwell (when he was a young man Orwell's name was Eric Blair). Huxley wrote to Orwell after the publication of his novel *1984*. This book would in time become one of the best-known books ever written. Huxley said in this letter to his pupil that he truly believed what was coming was pretty much what Orwell noted in his novel.

 DOI: 10.1201/9781003280668-6

I'll start painting for you the picture Huxley painted with the use of a metaphor.

6.1 CROME OR CHROME, HUXLEY'S YELLOW

When Huxley wrote *Crome Yellow*, he was 26. The story is set in a country house of Ottoline Cavendish-Bentnick (half-sister of the Duke of Portland) and her husband Philip Morell. Others have, since the publication of this novel, decided that the house resembles mostly the Garsington Manor, a country house in the village of Garsington in Oxfordshire. The owners of the house at the time when Huxley wrote his novel were members of the famous Bloomsbury Group. This was the set of intellectuals, writers, philosophers, artists, and a few stray people of not so clearly decided calling. The many contributions they made to English and more broadly European life of the era of varying nature – popular philosophers, writers, painters, you name it – have influenced intellectual and artistic life of the British Isles for some time. The visitors to this (the real place, not Huxley's invented one) manor included E. M. Forster, Virginia Woolf, John Maynard Keynes, D. H. Lawrence, Bertrand Russell, Siegfried Sassoon, T. S. Eliot, and others. Aldous Huxley visited too. His novel *Crome Yellow* is based on one such half-imaginary visit.

The visit is described purely from the viewpoint of this single visitor who is Huxley's alter ego. There are stories within stories that begin from the travel and arrival to the place. The arrival is greeted by the story of the previous owners, and so on. The threads are many, and, if we were to describe them in mathematical terms, they are more of a spiral nature than purely circular.

Now for something completely different. There is a colour called chrome yellow. Remember Vincent van Gogh's *Sunflower* series? The flowers in that painting have plenty of that colour. Ever seen the US school buses? They are decorated in that colour. A bright, saturated, vibrant colour that is derived from a lead chromate, a compound that was first synthesised in early 1800s. It is one of the "newer" colours. The younger you are, the more ridiculous that last sentence would sound to you, but just you wait.

6.2 SKETCHING WITH THE CHROME YELLOW

The chrome yellow is a bright colour, but it could be used for sketching. Dissolve it enough in oil and you can, with a broad brush, make a sweeping statement that you can then cover over with more precise moves. Or, like van Gogh, you can paste it in abundance to make your shapes come to life not only in colour but texture as well.

This beautiful little story of *Crome Yellow* alludes to chrome yellow. Major personalities include Scogan who apparently represents Bertrand Russell.

Russell (1872–1970) is the man who came up with the paradox now called after him, also after the Italian mathematician Ernst Zermelo (but Zermelo didn't make it to *Crome Yellow*). Russell's paradox is illustrated by the following little riddle, not necessarily invented by him (some say a friend suggested it to him), that goes like this:

> Consider a group of barbers who shave only those men who do not shave themselves. Then suppose that there is a barber who doesn't shave himself. What happens next? He should shave himself as he is a barber. But no barber in the collection can shave themselves.

Our story in this chapter is a story within a story – nestled in some way within another as in this one of Huxley's novels. The first will be something that seems unconnected to *Crome Yellow* but perhaps not so much. Enjoy the ride.

6.3 THE NUMBERS THAT KEEP COMING UP

Samuel Jeake said, towards the end of the 17th century, that perfect numbers are as rare as perfect men or women. But there are abundant and deficient numbers to make up for it! A, number is one of these three (perfect, abundant, deficient) as it is equal to, or smaller or larger, than the sum of its aliquot parts. The aliquot parts of a number are its factors, including 1.

> 6 is perfect as $6 = 1 + 2 + 3$
>
> 8 is deficient as the sum of its aliquot parts is smaller than the number itself, $8 > 1 + 2 + 4$
>
> And 12 is abundant as the sum its aliquot parts is larger than the number itself, $12 < 1 + 2 + 3 + 4 + 6$.

This has been known for a long time. Euclid (who flourished around 300 BC) proved the following in his *Elements* (published c. 300 BC):

> If as many numbers as we please beginning from a unit are set out continuously in double proportion until the sum of all becomes

prime, and if the sum multiplied into the last makes some number, then the product is perfect.

(EUCLID, IX:36)

This means that if

$$p = \sum_{0}^{n=p-1} 2^n$$

Let's translate this mathematical statement. p here is a prime number. Sign for sum is given in Greek sigma Σ. Then below this sign we have the value of n, which goes from 0, and above the sigma we can see it goes to the value of prime -1. And the sum is of all terms of a sequence where we look at 2^n, when n is taking values from 0 to $p - 1$.

In other words, if there is a prime such that the sum of a number of powers of 2 (beginning with the power of 0 which is = 1) is a prime number, then we will get a perfect number which will be equal to

$$\left(\sum_{0}^{n=p-1} 2^n \right) \cdot (2^p - 1)$$

This is nice to state, but it is more easily said than it is to find such numbers. Nicomachus found some of such perfect numbers. They were 6, 28, 496, and 8128. They all ended in 6 or 8.

Then Theon of Smyrna, father of Hypatia, confirmed only 6 and 28.

Iamblichus, another ancient Greek mathematician, asserted that there is only one perfect number in each of the intervals between 1, 10, 100, 1000, 10000, and so on. He also said that they will end in 6 and 8 only. We now know that neither of these statements is true, but you can find them repeated throughout the history of mathematics. For example, Boethius took this to be the case. It is believed that the authors from the Middle Ages and the Renaissance followed these two errors.

A little meditation on that perhaps is timely here. Then we ask ourselves:

Are there any odd perfect numbers?

Ancient Greek mathematician Euclid gave a method for constructing perfect numbers as earlier, but let's try a few for ourselves. Start with 1 then double.

$$1 + 2 + 4 = 7$$

7 is a prime, so we can use that. Now use 7 with the last number from the sum, which is 4 and multiply them

$$4 \times 7 = 28$$

Indeed 28 is perfect!

Now we can generalise a little. If you look at the original sum, you can rewrite it a little:

$$1 + 2 + 4 + \cdots 2^{p-1} = 2^p - 1$$

So now we can rewrite the whole formula from Euclid's algorithm to

$$2^{p-1}(2^p - 1)$$

Perfect. In more ways than one – a perfect algorithm and a perfect number. Now we can try to find the next perfect number by using the same method. This method is, by the way, called Euclid's algorithm for perfect numbers. Try to see how you get on. Although you know that they (perfect numbers) are so few and far between, you may be at it for a long time. And perfect numbers are as rare as perfect people.

Does that matter? Is the visit to the manor (from Huxley's story) any less important and not worthy of your time if you keep trying to meet someone that is very unlikely to meet a perfect human being?

Much after Euclid, Descartes and Mersenne corresponded on the subject of perfect numbers. Descartes once (in 1638) wrote to Mersenne to say that he thought he was able to prove that there are no even numbers which are perfect, apart from those generated by Euclid's algorithm. He also stated that he thought that there are no odd perfect numbers. The two are not the same:

- There are no even (perfect numbers) apart from those generated by Euclid's algorithm

- There are no odd perfect numbers altogether.

Both Euler and Fermat were also interested in perfect numbers. Euler found the eighth perfect number in 1732, and it is of the form from Euclid

$$2^{30}(2^{31} - 1) = 2305843008139952128$$

But no one has ever found any odd perfect number yet.

This is probably looking a little familiar. Perfect numbers are related to primes: if p is a prime, the sometimes you will find a prime (called Mersenne prime) of the form

$$2^p - 1$$

This is how we find biggest primes now days. The most recent one is $2^{136,279,841} - 1$, discovered on October 12, 2024. Wait a little, and there will be another. Another prime, another perfect number. If you wait long enough, you may find another perfect human. Usually, they appear when you give birth to them. Or when you fall in love. They are very rare to us individually, but there are infinitely many of them.

6.4 FURTHER EXPLORATION

Start from Huxley's *Crome Yellow* and do a little reading on his life. Then I would recommend reading his *Eyeless in Gaza*. Huxley, like Russell, was a pacifist.

CHAPTER 7

Prussian Blue and the Blue Waves

I SOMETIMES PAINT ON LARGE canvasses – I paint mainly just using various hues and shades of blue colours, and mix them in infinitely many ways (i.e. there is no description or a prescription on how to do that). Most often, I don't even bother with the shapes, just use colours. This does something good to my eyes and mind. I often wonder whether people would be happier if they concentrated on blue over any other colours. Out of all the blue colours that I mix and splatter and spread and tame with gentle movements of my brush, Prussian blue is definitely my favourite.

Prussian blue was the first synthetically created colour. Until then pigments were mined in nature. But with Prussian blue that practice changed. This colour was created in the early 18th century, and the story goes that it was discovered by Johann Jacob Diesbach in Berlin around 1706. Because it was born there, sometimes it is also called Berlin blue or Brandenburg blue, and sometimes Parisian or Paris blue. The way it was done is pretty orphic in a way: it was made from potash (which contains mined salts that contain potassium – the thing that is useful if you have raised blood pressure) mixed with blood to create cochineal dye. Now cochineal bugs are the bugs that also create ink – the dried one, that is. Huxley wrote about this too (in his *Eyeless in Gaza*), as well as about how perfumers make perfumes from certain things cats produce but no need to write about at the moment!

In any case, Prussian blue is a synthetic dye. It is significant in the history of pigments because it is stable (doesn't change hue over time), and since

DOI: 10.1201/9781003280668-7

we lost the knowledge of how to make Egyptian blue until recently, it is the one we could use for that same effect of blue – a colour that so strongly draws one in. Egyptian blue was used of course in Egypt and the knowledge of how to make it was passed on to the Romans. But for many centuries, Western civilisation had no idea how to make it – and this is where Prussian blue came useful when it was first invented. In fact, Prussian is much nicer than Egyptian blue – it is darker and deeper.

Now how this is connected to mathematics is a question of choice. We could investigate the frequency of these colours. But instead, we will look at one famous picture that is connected to Prussian blue and that also shows another mathematical motif that was not mathematically formulated for another century since it was painted.

7.1 THE WAVE AND THE FRACTALS

The picture is *The Wave* was painted in 1831 by Katsushika Hokusai. He was an old man when he did this, in his 70s. A famous painter in his earlier life, he thought he had retired, when his grandson gambled his fortune and Hokusai had to go back to work. The full title of this painting is *In the Hollow of a Wave off the Coast at Kanagawa* or sometimes it is translated as simply *The Great Wave off Kanagawa*. This is a hand-drawn diagram of the famous image of *The great wave off Kanagawa*, another little doodle.

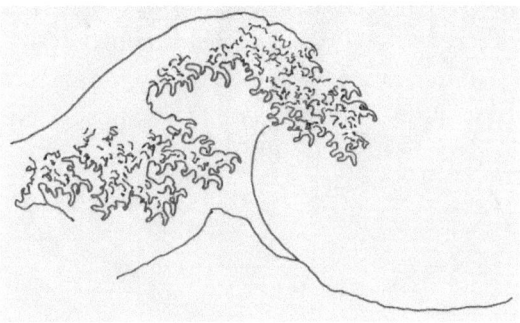

FIGURE 66   The great wave off Kanagawa.

This painting was not a one-off, it was part of a series of 36 paintings.

Like with any art, you can draw structures over it. You may have seen golden rectangles drawn over prints or reproductions of Mona Lisa (by da Vinci) at some point. We sometimes want to see what we believe is there. But *The Great Wave* is pretty close to something that is mathematical and that, in the time of Hokusai, hadn't yet been invented: fractals.

7.2 A GROUP OF MATHEMATICIANS OF ORDER THREE

French mathematicians of the early 20th century – so just about the time when Denis Stone, the main protagonist of *Crome Yellow*, visited the mansion of Crome – started thinking about mathematical structures that resemble this blue wave. In this story, the mathematicians were Pierre Fatou and Gaston Julia. Fatou was a student of the famous École Normale in Paris and friends with Maurice Fréchet. It was Fatou that created a study of analytic functions and introduced the set which is now called Julia set. This study is called *holomorphic dynamics* and deals with the dynamical system you get when you iterate a complex analytic mapping. This is somewhat of a mouthful, so let's see in plain terms what it refers to. A dynamical system is when you study a behaviour of some (mathematical) point dependent on time. Examples people put forward are the flow of water in a pipe or pendulum of a clock.

Let's take that pendulum of a clock. I won't ask you to imagine it and hypnotise you unintentionally. But thinking of a pendulum clock is a nice way of meditating on the dynamical systems. Now, you can obtain a mathematical model of such a system by iterating a complex mapping. Iteration is a recursive, repetitive process where every next time you do it, you get some kind of approximation to the better or worse of what you want to see. And complex analysis deals with looking at functions of complex variables – complex numbers.

A complex number is a number such as one that is represented by a diagram where there is a real part and an imaginary part to the complex number. The real numbers are plotted on the horizontal, x axis. The imaginary part of such a number is plotted on the y, vertical axis.

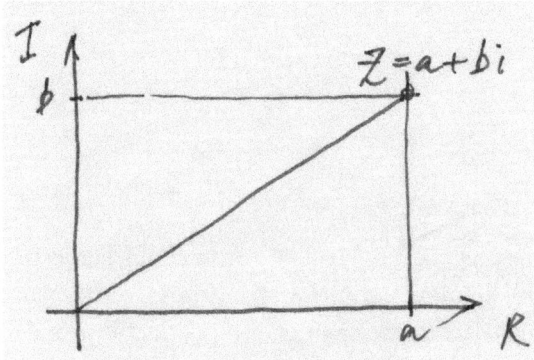

FIGURE 67 A complex number diagram – real part on horizontal axis, and imaginary on vertical axis.

Gaston Julia was a French mathematician, and the one who devised the formula for the Julia set. That's why it is called after him and not Fatou, although Fatou introduced the study of this set in the first place. Julia was an Algerian-French mathematician.

The Julia set and the Fatou set (as Fatou had one named after him too) are complementary. The Julia set would look like this,

FIGURE 68   Simplified Julia set fractal.

and the Fatou set is usually a little more boring. The thing is that the Fatou set of the function in question has values which, when you iterate them repeatedly, behave very similarly to the original. But the Julia set has such values that with small changes in iterations, you can have really drastic changes so much so that its behaviour is called *chaotic*.

7.3 WHAT DOES B IN BENOIT B. MANDELBROT STAND FOR?

Mandelbrot comes to mind now. Benoit B. Mandelbrot was also French, Polish born, and became an American mathematician. With the big migrations of the 20th century initiated with the rise of the Nazis, Mandelbrot emigrated from Warsaw to France at age 11. He studied in Paris but, being Jewish, had to hide with family during WWII. After the war, he studied under Gaston Julia and Paul Lévy, also a friend of Fréchet. After WWII he moved to Switzerland, then back to France, and finally to the US, where he remained, working for IBM – International Business Machines Corporation.

This corporation was named *Big Blue*, but as much as I would prefer it, it was not the Prussian blue (more like Egyptian blue) that was employed in the making of their logo.

Not unsurprisingly, Benoit was fascinated by the sudden and chaotic changes. How could one avoid searching for answers after something like what happened to him and his family during WWII took place? He wanted to understand some of the aspects of such chaotic behaviour through mathematics. If not necessarily for himself any longer, as the world entered a relatively peaceful period after 1945, but for other children, and for their futures.

In later life, Benoit realised that financial markets had something in common with what was happening with societies during the times when critical changes were taking place. So he looked at these fluctuations. Afterwards, he tried to find out something more of mathematics from the time he studied mathematics under Gaston Julia.

He came across Julia sets and now he was able to use computers as he was working at IBM. While he studied these sets, he came up with another set, which is now called the Mandelbrot set. It was Benoit that coined the term *fractal* to describe these structures. He described them as the *art of roughness* or models of those phenomena that can be experienced as *uncontrolled element in life*. As such, it is good that everyone should study this to a certain degree.

Let's start very slowly and see how iterations can be stable and how they can turn violently chaotic. By the way, to make a fractal means making iterations that repeat what is already there. In Benoit B. Mandelbrot, you start from B and go to Benoit . . . B. Mandelbrot. And so on.

7.4 FRACTAL MAKING

What's a fractal? Let's start by looking at how iterations can provide completely different results based on their *seeds*. You remember the labyrinth seeds – this is equivalent, as it is a *base*, or the original thing that you will use to repeat in different ways to create a more complex structure.

You start with a certain value or a shape, and you get the next iteration by using that value *and* an algorithm to get the second value. And you go repeating the process with the resulting value as the new input value, and go through the same algorithm over and over again.

We'll do the simple one as this is one of the only ones that we can draw easily without resorting to computers. If you want to do it on your own, this is the one to play with. Iterations here we come. Our function is x and it maps into x^2 until you tell it to stop.

Try some values.

1, then square that, and it becomes 1, and square that, it becomes 1, and then 1, you get the gist. This is not an exciting iteration due to the seed we have chosen.

The important thing to realise is that seeds also need to be interesting. Let's now try 5. We get (let's say for the four steps and you'll see why so few very soon):

$$5 \rightarrow 25 \rightarrow 625 \rightarrow 390625 \rightarrow 152587890625$$

The values escalate very quickly!

Now let's try with a negative value. Say −3.

$$-3 \rightarrow 9 \rightarrow 81 \rightarrow 6561 \rightarrow 43046721$$

This still goes large pretty soon. You can try with numbers smaller than 1, and that will go the other way round

$$\frac{1}{2} \rightarrow \frac{1}{4} \rightarrow \frac{1}{16} \rightarrow \frac{1}{2561} \rightarrow \frac{1}{65536}$$

Now the number becomes small(er) very quickly too. But how much smaller? Are they going to go into some negative value? Is the small we are talking about here going towards $-\infty$? Of course not, as we are squaring every iteration. So even if we start from a negative, once we square it, it will become positive (minus × minus = plus). So the *small* we are talking about is the value approaching 0. But never quite reaching it.

Now we can go on with this meditation, but it would require either a good knowledge of complex numbers or access to a computer. By changing the constant which we iterate, we can get quite different shapes of fractals, of Julia and Mandelbrot sets. You can find a bit about mathematicians mentioned here at the end of the book. Doodling one of these may be fun, but it won't be precise enough (mathematically speaking) for you to enjoy that too much. But you can certainly try, and establish some kind of rule to repeat over and over again.

The fractals are a little more difficult to do and definitely difficult to doodle. We'll instead do something that is easy and no less exciting and interesting. And much more precise. Let us do a hand-drawn diagram of the Sierpiński's triangle.

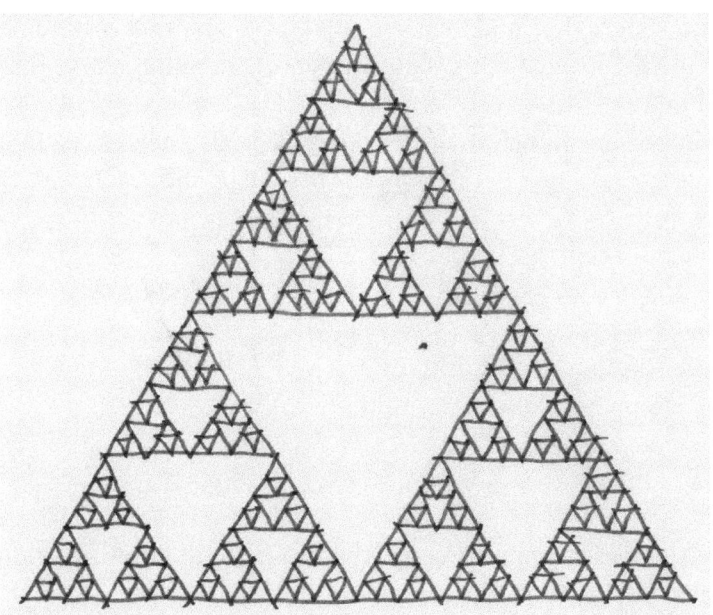

FIGURE 69   Fractal – Sierpiński's triangle.

You would have seen this before, the triangle that is called Sierpiński's. It is named after a Polish mathematician Wacław Franciszek Sierpiński (1882–1969) who made major contributions to set theory and number theory as well as topology. A prolific writer of mathematical works, he left behind him about 700 papers, founded a major mathematical journal, and published around 50 books. He was busy all his life.

And you can get busy making this fractal. You can start imagining how to create this or even doodle it. The triangle is an equilateral one.

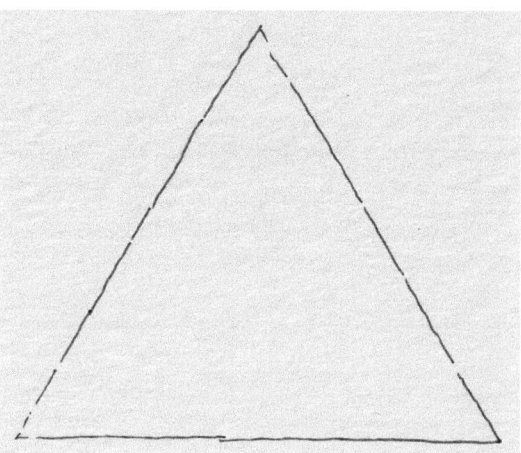

FIGURE 70   An equilateral triangle is all we need to start this fractal.

Now you need to find the middle of each side of the triangle and connect them. This is the first iteration of the Sierpiński's triangle – a triangle with an inscribed triangle connecting mid-points of the original.

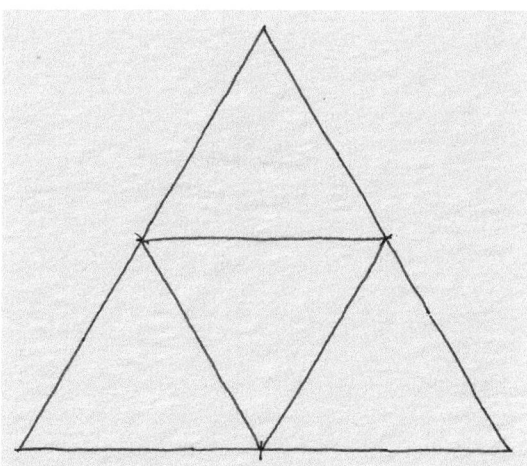

FIGURE 71   First iteration of the Sierpiński triangle.

You can now make another set of triangles in the same way in each of the smaller triangles. But let's keep the middle one empty. And whenever we get to the next level of iteration, let's always keep the middle one empty. The next step would look like this: a new triangle appears with an inscribed triangle connecting mid-points of the original.

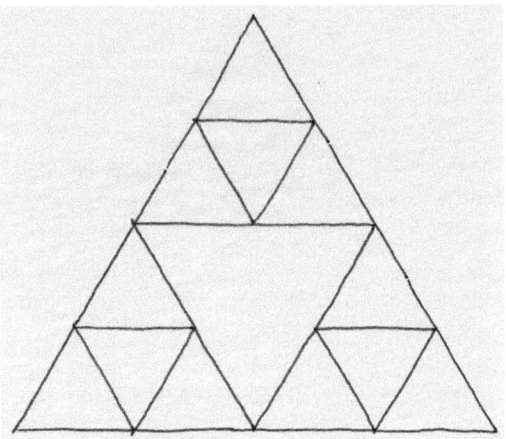

FIGURE 72 Second iteration of Sierpiński triangle.

Now you can keep doing this – keep having further iterations in the same manner. Keep the middle always free from further iterations. Find midpoints of the other three triangles and construct new smaller triangles within them.

FIGURE 73 Third iteration of Sierpiński triangle.

You can continue as long as you like and the size of your drawing allows you. There is a strange satisfaction in seeing order of greater and greater number of smaller triangles as you go along. You can even look at a whole individual set of iteration as you make them individually. The next one would look like this.

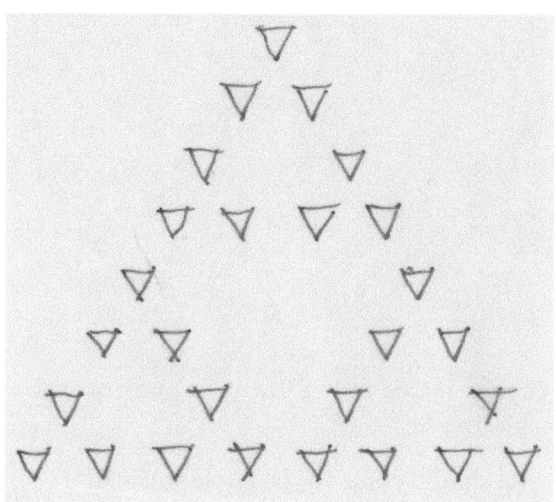

FIGURE 74   A layer of iteration on a separate sheet of paper.

In case you are imagining this, you can certainly get to making a great amount of detail in your mind, and doing many more of these iterations than you could do in reality. This is usually the case – imagination can take you to places where you can't necessarily get to in reality. In that case, what you will notice is that in order to imagine how to do this, you will *zoom in* to your smaller and smaller triangles to see their next iterations.

This brings us to the first way to describe fractals in general – their *self-similarity*. The shapes are always the same, or rather, similar – same in shape but smaller or larger. You may want to do your meditations on creating fractals from other shapes. What about circles? Inside a circle you need to first decide which kind of structure you will follow and then continue doing the same thing in iterative process to your smaller circles. Perhaps work out what I have done here and do a doodle of your own. Here, a circular shaped fractal is created with iterations based on the division of circles in four equal arcs, and then by construction of inverted quarter circles inside it.

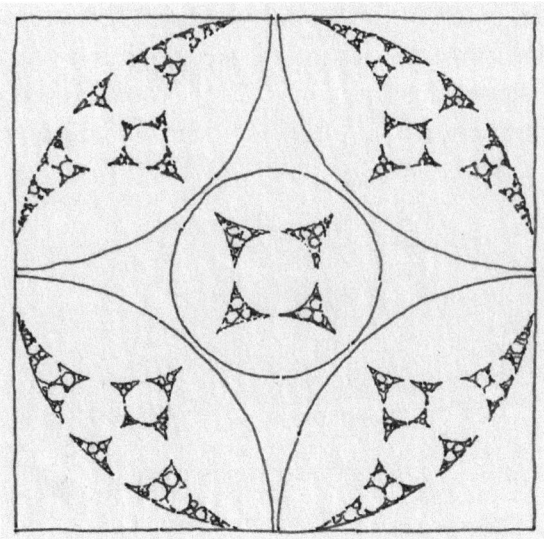

FIGURE 75   A circular shaped fractal. There is a similarity between this one and Sierpiński's triangle. In which way?

Eventually, you should come to the circle so small if you do a fractal as this one, it will look like a point. That brings us back to the point of doing iterations with our imaginary stories. The best short story of all times for me is the *Library of Babel* by Borges. It starts as follows . . .

7.5 THE IDEAL LIBRARY

The universe (which others call the Library) is composed of an indefinite, perhaps infinite number of hexagonal galleries. In the centre of each gallery is a ventilation shaft, bounded by a low railing. From any hexagon one can see the floors above and below – one after another, endlessly. The arrangement of the galleries is always the same: Twenty bookshelves, five to each side, line four of the hexagon's six sides; the height of the bookshelves, floor to ceiling, is hardly greater than the height of a normal librarian. One of the hexagon's free sides opens onto a narrow sort of vestibule which in turn opens onto another gallery, identical to the first – identical in fact to all.

(Opening lines from the story, published in 1941, by
Jorge Francisco Isidoro Luis Borges Acevedo – 1899–1986)

Now you can begin drawing this structure, either as a plan or imagine it. In that case it may appear in deluxe colour while you meditate. Some have tried to recreate the floor plans from the story, which is in places contradictory. Hence there are a few different possibilities, but here is one diagram to start from.

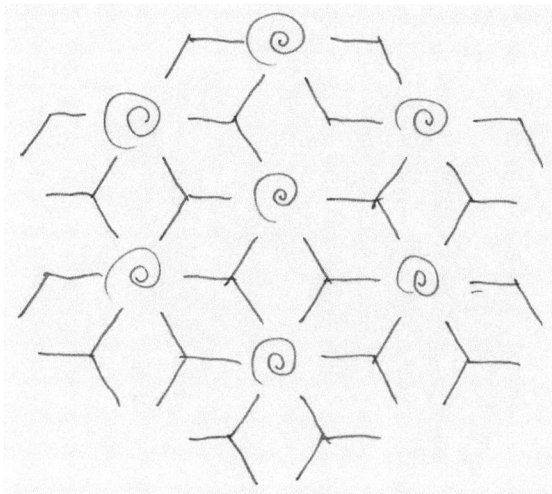

FIGURE 76　A doodle of the possible plan of Borges' Library.

The hexagonal cells are lined with books; they are connected to each other by passages, and in between them every so often there is a spiral staircase leading to other rooms in the library. Where would mathematical books be kept? Amongst all others or in a special place? As you imagine picking the books you remember to remind yourself of the experiences you had for the first time when you read something. Bertrand Russell mentioned once that reading Euclid's *Elements* that his brother gave him was like falling in love for the first time.

Then, on the other hand, you may want to create a floor plan of your ideal library and trace your journey through it. How many such galleries would you want to cover in how much time you have? What books would you imagine on the shelves of this library?

Libraries are often organised in a way to support easy retrieval and memorable arrangement that would allow you to find your ways around them easier.

There is an ancient technique so well described in a book by 20th-century art and science historian Frances Yates, *The Art of Memory* (1966). The art of memory has many techniques that could be employed in pursuit of memorising but also classifying and reflecting on what has been learnt and read. There is one technique I recommend above others. It is to imagine a room, and clarify for yourself all its arrangements and relationships between things and structure of that room. Then try to remember everything about it and in it. You may construct a room entirely in your mind. I like to imagine my study in this exercise. I know exactly (within its reappearing chaos) where things are. The size of the room, the position of the window, the sizes and content of my bookshelves; the position of chair, desk, and my little objects that remind me of the dearest people from my life. Anything that takes your fancy and you can easily use to jog your memory is good.

So now you have made a decision – it could be your study, or it could be one of the cells of Borges' library – in it you should place your important books and objects. Remember where you have put what, and look at things occasionally in your mind's eye. When you come back to this room and pick a book from its shelves, you will find a memory that is important to you.

FIGURE 77   A memory room may be a memory theatre (as from Yates' book).

7.6 FURTHER EXPLORATION

You may now decide to create an imaginary library where you can place an infinite number of interesting books and papers. This time, therefore, here are a slightly larger number of things you may want to explore further.

A book on the history of *The Art of Memory* by Frances Yates was published in 1966 by Routledge and Kegan Paul.

The *Library of Babel* is a short story by Borges and can be found online in many languages. It was first written in 1941.

There is a whole book dedicated to a kind of mathematical meditation based on this short story by Borges, under the title *The Unimaginable Library of Babel*, written by William Goldbloom Cohn, published by Oxford University Press, 2008.

CHAPTER 8

Que Sera, Sera (What Will Be Will Be)

L ET US IMAGINE A meeting, not of the kind where a philosopher and two mathematicians walk into a bar, and certainly not at the same time. It is a meeting of a sort nonetheless. The three are Cicero (1st century BC), Kepler (1571–1630), and Grothendieck (1928–2014). They are not the only ones in the history of philosophy who grappled with this question, but they are interesting, and particularly interesting as we contemplate the mathematics that is weaved into their lives' stories.

8.1 THE DREAMS AND DREAMERS

Cicero (106–43 BC) was a sceptic, which means he doubted that total knowledge was ever possible. He was not a mathematician, but in one of his writings, in a portion of his final book, *De re publica*, there is a segment that is interesting for us. This was written around 54–51 BC. Cicero wrote about the type of government that existed in Rome and the challenges posed by Caesar. The portion of this work is his *Dream of Scipio*, in a way a science fiction novel about space (and time) travel. Scipio Aemilianus was a Roman general, and his dream, described by Cicero, took place some two years before he oversaw the sacking of Carthage in 146 BC. As he arrives to Africa, he falls asleep and is visited by his grandfather, Scipio Africanus. His grandfather tells him about his future – and reminds him that it is important to uphold loyalty as well as rightful conduct regardless of what comes to him next. Scipio Aemilianus finds himself looking from above onto the

 DOI: 10.1201/9781003280668-8

Earth. He realises that Rome is quite insignificant in the grand scheme of things (in the universe). However significant it seemed to him when he was awake, in all the cosmos it is not that great. He looks at the space further from the planet Earth and realises that it is made of nine celestial spheres. The Earth is in the centre, and all the rest surround it – the heaven is the highest, containing everything else, including the divinity.

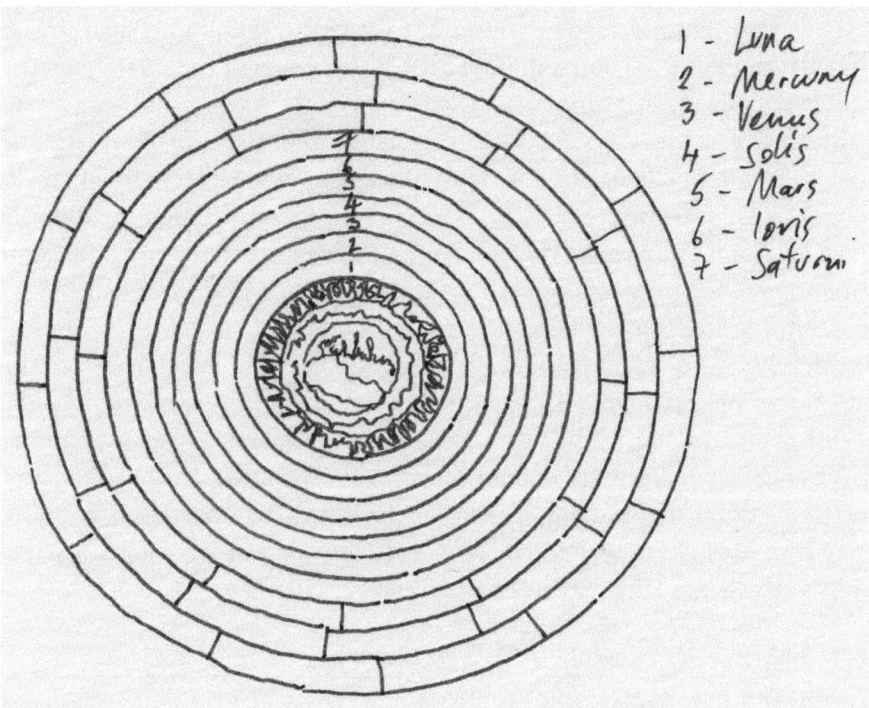

FIGURE 78   A doodle based on Peter Apian's illustration on geocentric celestial system (1539).

Of course, this is a geocentric model of the universe that we know is not correct, but that is not as important as what follows. Scipio, the dreamer, sees that between Earth and heavens lie seven spheres beginning with the Moon, then Mercury, Venus, the Sun, Mars, Jupiter, and Saturn. As he is amazed with that, he is further amazed by the sweet and great sound between them – the music of the spheres – the *musica universalis*.

Many people have seen and read this text since it was written. Did Kepler read this? It would be a bit of a coincidence if Kepler wrote a similar story without reading *The Dream of Scipio*.

Johannes Kepler (1571–1630) attended Latin school and studied theology initially. He certainly recreated a number of things from Cicero's dream. Kepler wrote a little book, *Somnium* (*Dream*), about a similar event. Only in Kepler's dream the traveller to the Moon was not a general but a scholar scientist, just like Kepler himself – and he was led similarly to see Earth from afar. He travelled to the Moon in this *dream*. Kepler also thought about the spheres that structure the known universe, only he did not put the Earth in the centre of his universe as by this time Copernicus (1473–1543) had already shown that it was the Sun and not Earth in the centre of the solar system.

Later, in his *Mysterium Cosmographicum* (1596) Kepler showed a model of the universe based on Platonic solids, encased in a sphere, repeated until they were all included. This he eventually realised, years later, wasn't really how the universe was structured, but it was a useful model to start his investigations from. Now that we have Kepler's laws of planetary motion, this model is a relic of Kepler's original and erroneous thinking, but a beautiful example of such thinking, nonetheless.

The then known six planets of the solar system Kepler included in this system of nested solids. They each fit in their spheres, Mercury, Venus, Earth, Mars, Jupiter, and Saturn. Kepler is said to have started working on this model when he learnt about the *musica universalis* – the harmony that exists between the spacing of the planets and their movements in space. This divine music was believed to be heard only on rare occasions – maybe only at the beginning, when the universe was born?

8.2 HARMONY OF THE WORLD

Cicero and Kepler, divided by centuries, had described similar "dreams" of extraterrestrial travel from which they gained some knowledge that would not have been possible to describe otherwise. Did they actually dream these scenes themselves or just imagined such travel? Considering that we, as a species, have been able to travel to the moon centuries after Kepler, his *Dream* is also a foreknowledge of what was about to happen.

Two *dreams* differ in their models of the universe. Cicero's is geocentric, Kepler's Copernican system is heliocentric. Then there is a mathematician closer (historically) to us, Alexander Grothendieck (1928–2014). He didn't need to talk about a model of the universe. He was talking about something more ephemeral.

Grothendieck was a famous mathematician and became an influential one both in France and globally. But he became a recluse and withdrew from public life half-way through his life. People have conjectured on the reasons for this, but this is not of concern for us here. He also wrote about a "dreamer" towards the end of his life. His writings on this are found in a manuscript that is similarly devoted to dreaming, entitled *La Clef de Songes* (*The Interpretation of Dreams*).

There is a painting of René Magritte, with the same title – *La Clef de Songes*. You can imagine a version referring to mathematics, a version all of your own. Here is mine, a hand-drawn dodecahedron with a sign underneath it.

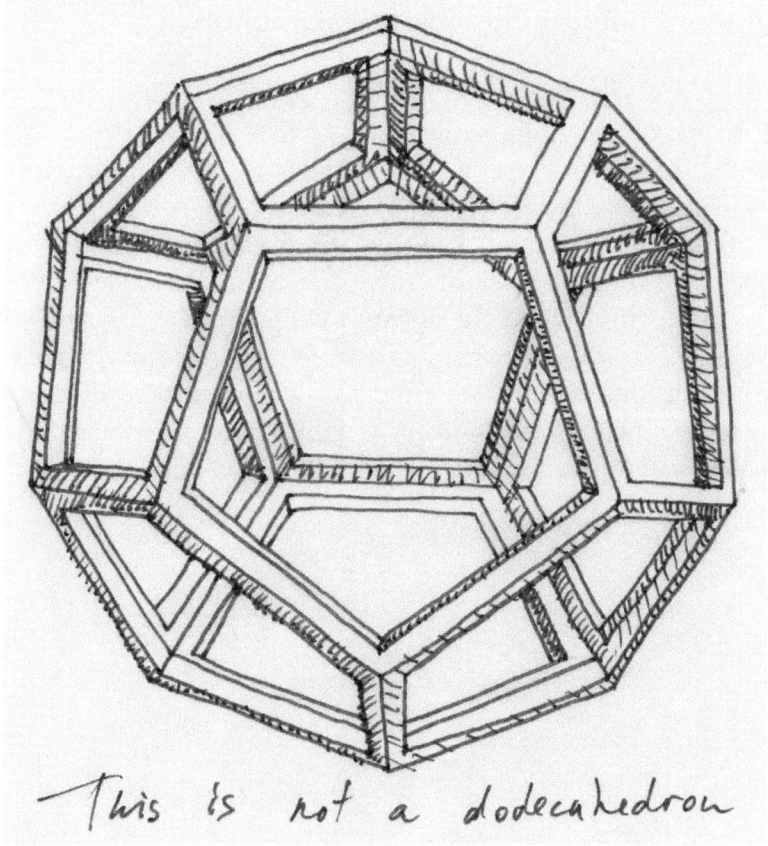

FIGURE 79   Les clé des songes mathématiques.

For Grothendieck, dreams are sent by the *dreamer*. This dreamer is a divine being of a kind. The dreamer sends us dreams so that we can learn about ourselves. Although in this capacity dreams are a message from the dreamer, there is a learning that happens in dreaming, just like with mathematics. At the time that Grothendieck wrote this, he was already shunned by the established academia and lived as a hermit in the Pyrenees. The role of mathematics in his opinion was to alert us to this whilst we are awake. Mathematics can be like a dream, a meditation on truths that we are given otherwise, by the *dreamer*.

Both Cicero's treatise and Kepler's dreams, as well as Grothendieck's key to dreams, are inspired by an idea that is in its core mathematical. There is a harmony of the spheres and a divine music emanating in space. Mathematics is used for us to understand it. Whilst keeping in mind that this is all a mathematical meditative game of a kind, let's go through some original steps to focus on some old beautiful mathematics.

8.3 BEAUTIFUL SOLIDS

How to start untangling mathematics of this? The spheres of Kepler's system of the universe rest on Platonic solids. So let us begin from these mathematical objects.

Platonic solids are those described by Plato in *Timaeus* (c. 360 BC). These are the only five regular solids in three dimensions. Plato identified these with elements from which the universe was built: cube was representing earth, octahedron air, icosahedron water, tetrahedron fire. Platonic solids have the same number of faces and the same number of edges meeting at every vertex: cube has three squares and three edges; dodecahedron three pentagons and three edges and so on.

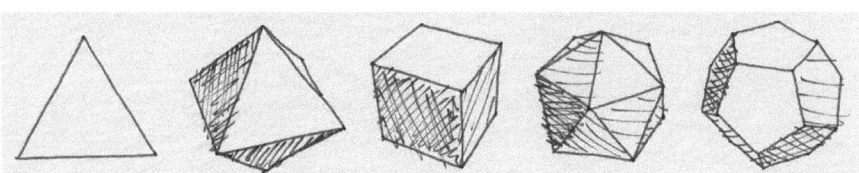

FIGURE 80   Five Platonic solids.

In order in which they appear in the image, tetrahedron would represent fire; octahedron – air; cube – earth; icosahedron – water. But there are five of those – what does dodecahedron represent? A divine spark, a principle of life, it is the ultimate element without which none would come to life, and from which there wouldn't be a universe at all.

The divine spark can be interpreted in many ways, one of which is our urging desire to love but also to learn, to find something out, and to find meaning in things we experience. Such researches reached their peak in Kepler's other work, *Harmonices Mundi* (*Harmony of the World*).

Harmonices is divided into five books, the first two of which present a study in Platonic and Archimedean solids. Since antiquity, no one has ever drawn the Archimedean solids – the semi-regular polyhedra. Kepler gave their depictions in his book 2 of the *Harmonices Mundi*, before going on to talk about his theory of the divine music of the cosmos. Let us then recreate them here just for the enjoyment of doing so.

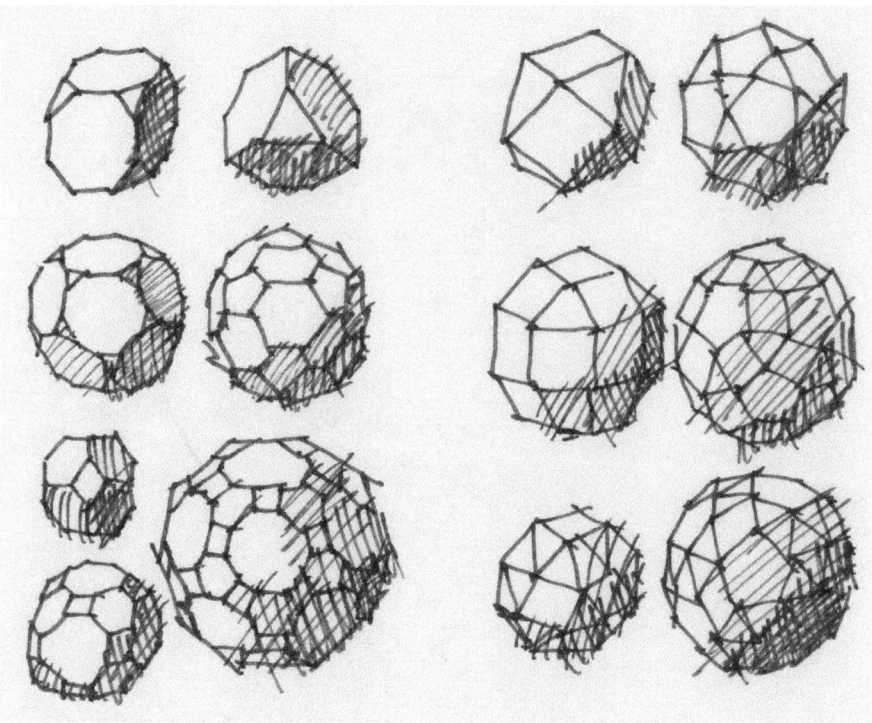

FIGURE 81   Archimedean polyhedra (from top left to bottom right in two groups).

There are 13 of these – called after Archimedes (c. 287 BC–c. 212 BC). Their main characteristic is that each face is entirely visible on the outside of the solid, and each vertex is surrounded by regular polygons arranged in the same way. On the image, they are in two groups – the left is of truncated solids, and the right is of combined ones.

Let us recreate one of them, the cuboctahedron, just to test the statement by Grothendieck that mathematics somehow gets you into a dreamland of a kind.

This Archimedean solid was recreated in the Renaissance, and before Kepler, by a Florentine painter-mathematician Piero della Francesca (1415–1492). He described it and constructed it in his *Trattato d'abaco*, which was not published for centuries, but was written somewhere around 1460. Let us do a doodle ourselves.

Start from a cube.

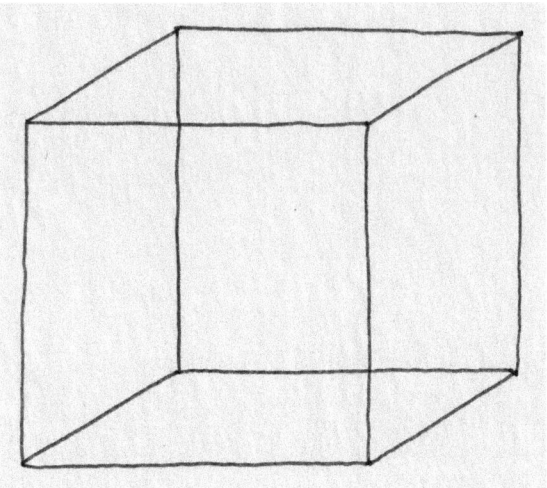

FIGURE 82   A cube.

Find midpoints of all the edges.

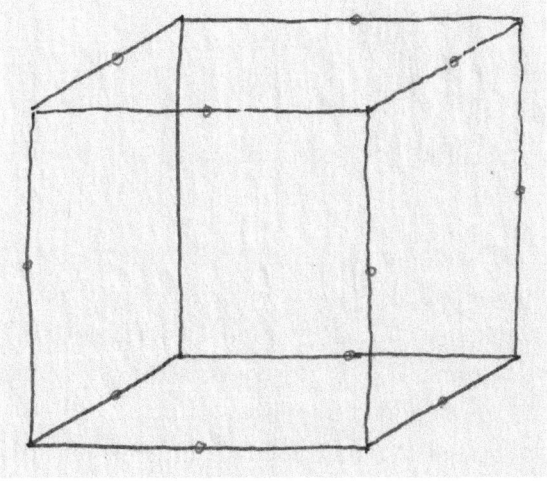

FIGURE 83   A cube with midpoints on edges.

Let these midpoints be vertices of your cuboctahedron.

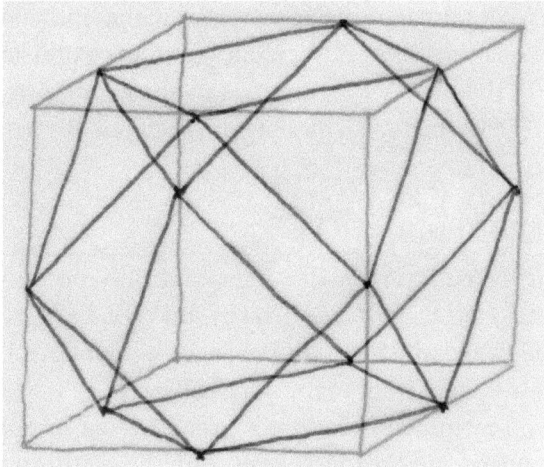

FIGURE 84   You will get a cuboctahedron inscribed in a cube.

And you can tidy it up a bit, to see it better, count its faces, edges, and vertices, and see what kind of faces it has. You can even calculate, if you wish, the areas of faces – in what ratio would the different shapes' areas be to each other? And you can make a little picture out of it too.

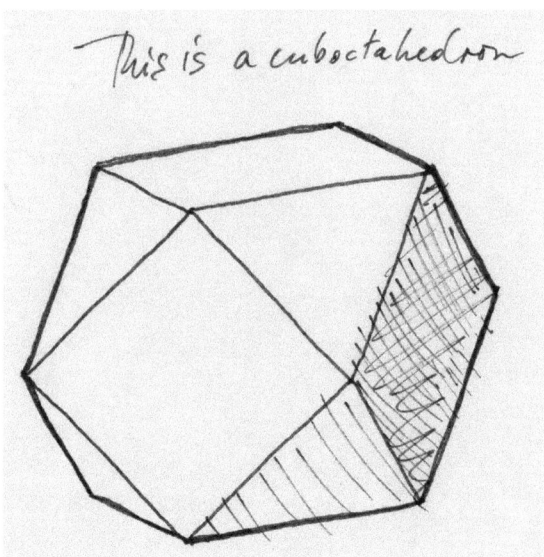

FIGURE 85   This is cuboctahedron.

If you have now done this – recreated a cuboctahedron like Piero did it, then you would have, I hope, found some pleasure in the task. Maybe on this one aspect of mathematics you will agree with Grothendieck. For him, there are clear relationships between our psychical world and the sense of apprehension of the beauty of mathematics through doing mathematical exercises with whatever tools you may have at your disposal. A pencil to make a doodle will often do.

8.4 FURTHER EXPLORATION

You may want to see a very pricey but beautifully produced book *Piero Della Francesca: A Mathematician's Art* by Judith V. Field. It was published in 2005 by Yale University Press and soon went out of print. Second-hand copies are rare and hence not easily obtainable.

I would also recommend reading Kepler's *Dream* or *Somnium*, which has been republished in various places. For those who want something to read for and with children, *Kepler's Dream* by Juliet Bell is a wonderful little story.

CHAPTER 9

Questions and Answers

There was a child went forth every day

There was a child went forth every day,

And the first object he looked upon and received with wonder or pity or love or dread, that object he became,

And that object became part of him for the day or a certain part of the day . . . or for many years or stretching cycles of years.

The early lilacs became part of this child,

And grass, and white and red morningglories, and white and red clover, and the song of the phœbe-bird,

And the March-born lambs, and the sow's pink-faint litter, and the mare's foal, and the cow's calf, and the noisy brood of the barn-yard or by the mire of the pond-side . . . and the fish suspending themselves so curiously below there . . . and the beautiful curious liquid . . . and the water-plants with their graceful flat heads . . . all became part of him.

Walt Whitman (1819–1892)

E VERY QUESTION HAS SOME purpose of finding things out. If you want to know a good answer, you need to ask a good question. But sometimes you will not find the answer, or it will elude you. Some questions cannot be answered by even the most capable mathematicians. Reasons could be different for different people and for different questions. Perhaps mathematical tools haven't yet been invented that one could use to solve the problems which you put in the centre of your question.

DOI: 10.1201/9781003280668-9

There are some very difficult problems for professional mathematicians, and we will leave those aside. But, if you are looking for some number puzzles that are difficult enough for you to really get stuck in, but simple enough to work with children, read on. You can try these with kids and ramble through the number-land to your heart's content. The questions here are not as serious as those a child from Whitman's poem may ask. Or, who knows?

9.1 GOLDBACH CONJECTURE

Christian Goldbach (1690–1764) was born in the city of Königsberg, which we already visited at the beginning of this book. He was the best friend with Leonhard Euler, and from him Euler learnt of the bridges problem we mentioned in Chapter 2. They became friends as Goldbach also went to St Petersburg, when Peter the Great established the new university there, and the Russian Academy of Sciences in 1725. He (Goldbach) even became a personal tutor to Peter II. He knew many mathematicians around Europe.

In 1742, at the time when Euler was living in Berlin, Goldbach wrote him a letter. The letter stated a conjecture: every integer greater than 2 can be represented as a sum of two prime numbers. Euler responded by saying:

> There is little doubt that this result is true . . . that every even number is a sum of two primes, I consider [this] an entirely certain theorem in spite of that I am not able to demonstrate it.

But later, in the 20th century, Russian mathematician Ivan Matveevich Vinogradov (1891–1983) showed that if we look at *sufficiently* large odd integers, we can write them as the sum of at most three primes. From this followed that every such sufficiently large integer (not necessarily odd) is the sum of at most four primes. One result of Vinogradov's work is that we take Goldbach's conjecture to hold true for *almost* all even integers.

Try for yourself, you may come across a counter-example. If you do, you would have proved that this is not case. But let's do a few sums.

Get any even number such as 8. It can be written as $5 + 3 = 8$. Try some other, say 20. It can be written as $13 + 7 = 20$.

What can you do with this? As it hasn't been proven or disproven, you can play with it. Two mathematicians Adam Cunningham and John Ringland from State University of New York, Buffalo, started playing with this. They wrote primes on both sides of a number triangle and then added some to see what they would get.

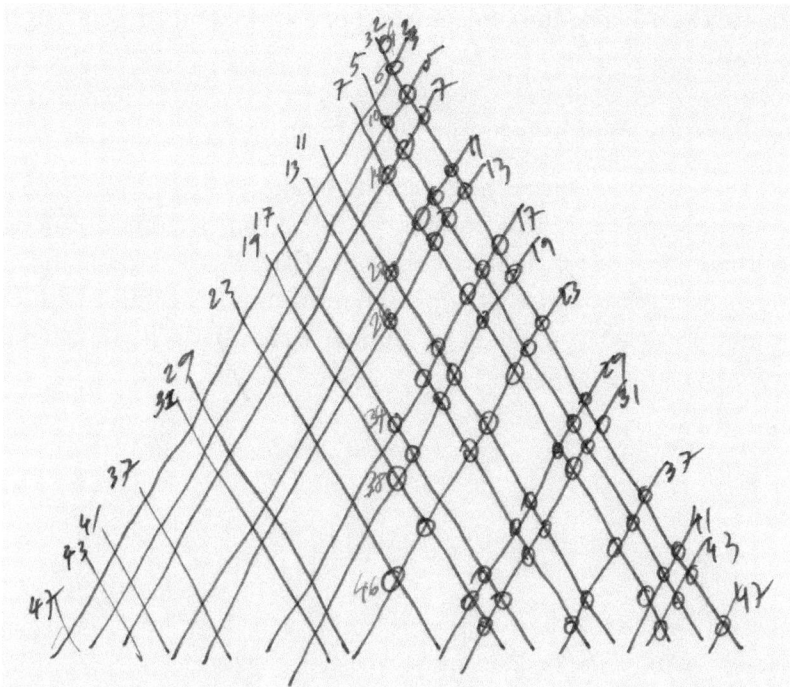

FIGURE 86   Trying to solve Goldbach's conjecture.

They did the same for even integers to 500, and who knows maybe more! You can try to beat them.

Goldbach's conjecture hasn't yet been proven or disproven.

9.2 TWIN PRIME CONJECTURE

Primes are numbers that can be divided only by themselves and one, and if you want to have twins between them, they would have to be of a difference of two between them. Twin primes then, among the small ones, are

3 and 5

5 and 7

11 and 13

17 and 19

29 and 31

41 and 43

There are two twin prime conjectures (not a surprise really, considering). They state:

a. There are an infinite number of twin primes . . .

b. The scarcity of the twin primes can be expressed by a number called Brun's constant. This constant number comes from an important theorem called Brun's theorem, which says that if you add the reciprocals of the odd twin primes, ad infinitum (to infinity, without an end in sight), you will get a number which will tell you how many twin primes there are.

We can write a Brun's constant formula therefore:

$$B \equiv \left(\frac{1}{3} + \frac{1}{5}\right) + \left(\frac{1}{5} + \frac{1}{7}\right) + \left(\frac{1}{11} + \frac{1}{13}\right) \cdots$$

Neither of these two conjectures are confirmed. It is, however, accepted that every twin prime greater than three is of the form

$$(6n - 1, 6n + 1)$$

for some natural number n. We will adopt a convention that the set of natural numbers doesn't include 0. More often than not, mathematicians would not include 0 in natural numbers as the preferred definition is that natural numbers are positive integers. But computer scientists would often include 0 for the reasons of sets that are important in computer science. Nevertheless, in our prior conjecture, having 0 would make the formulae obsolete.

So let's try a few:

5 and 7 – that's a check

11 and 13 – that's a check too

17 and 19 – works too.

In practice, these are found by computer trawling and so as of August 2022 the largest twin primes are

$$1996863034895 \cdot 2^{1290\,000} \pm 1$$

Now you can't really play with this unless you have access to powerful computers. But you can play with smaller twin primes and their properties. For example, every third odd number is divisible by 3. Let's try with 3 which is the first odd number. The second next one from it would be 5, and the next after that 7. The next one after that would be 9, which is divisible by 3.

As a consequence, no three successive odd numbers can be prime unless one of them is 3 (as others will be divisible by it! So not prime). As the only even prime is 2, our twins have to be both odd. Let's look at 5. Funny thing is, it appears in two twin pairs: 3 and 5, and 5 and 7. But that will not happen again.

You could play with this further. A list of the first twin primes up to 1,000 is given at the end of this chapter so you can play without going through the list to spoil your fun. Are there any other rules or patterns you can notice?

9.3 196-ALGORITHM

Numbers are curious and wonderful mathematical objects. Some have really curious properties. Sometimes you may think as you look through a row of numbers that they are a little boring. But then you can think if they are so dull, that makes them interesting, like the ones we just looked at.

Now let's look at just the dull numbers. There must be the smallest of the dull numbers. Wouldn't that be an interesting one? And what would be the largest dull number? Even probably more interesting. By using this type of thinking, you inadvertently stumbled on the proof by contradiction. And proved that there really is no number that is really dull.

But let's try our luck. Here's a number that looks pretty dull – 196. We are going to use it to make a particular type of numbers. The numbers we want to make are called palindromic. They are numbers that are self-symmetric as they can be written in both directions in the same way. For example, 1551 is a palindrome, so is 166661.

Let's try some palindromes. For example start with

5280

Now reverse it

825

(0 in the front doesn't make a difference, so no need to write it).

add the two:

$$5280 + 825 = 6105$$

Now repeat the process:

6105

reverse:

5016

add the two:

$$6105 + 5016 = 11121$$

Repeat the process again:

11121

reverse:

12111

add the two

$$11121 + 12111 = 23232$$

and finally!!

23232 is a **palindrome**!

You can try this with various other numbers.

The 196-Algorithm is called as such because 196 is one of the numbers on which this procedure *does not* work! Since the 1990s people have tried to find the palindrome in a number sequence produced by the aforementioned procedure, when it began with 196. Tim Irvin produced a number having two million digits by the procedure (starting from 196), but has not reached a palindrome.

Talking of palindromes, perhaps you have heard of the magic square with palindromic words excavated from the ancient Pompeii? It spells *Sator Arepo Tenet Opera Rotas*, meaning something like *the famer Arepo keeps the world rolling*, although there are other interpretations.

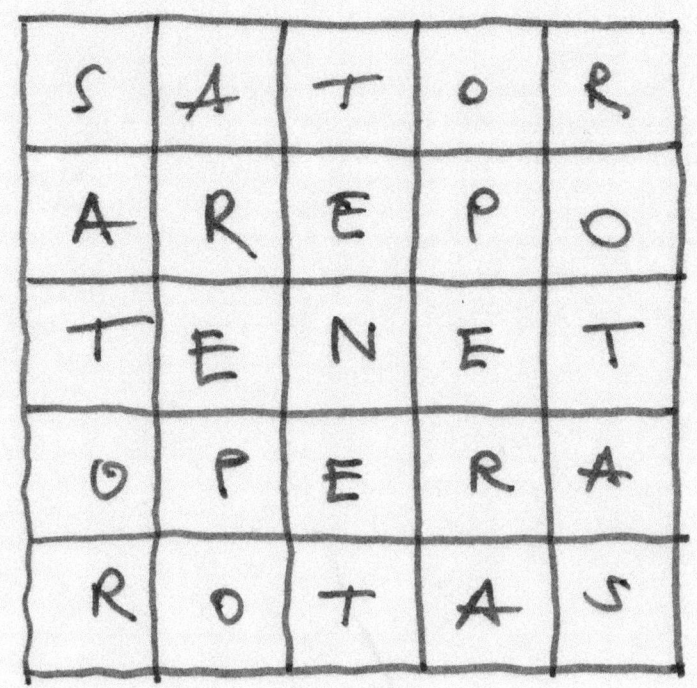

FIGURE 87   A Latin palindrome, found in Pompeii.

It is a magic square without any numbers (although you can substitute letters for numbers just like you can substitute numbers for letters). Every way you look at it, it will spell the same, and the middle cross would say *tenet*. In English it is about holding a belief, in Latin pretty much the same although it could be about anything that is held (opinion for example).

9.4 IS 10 A SOLITARY NUMBER?

There are friendly numbers and then there are solitary ones. A friendly, or amicable, number is the one which is part of a friendly pair. Friendly pairs are numbers that have divisors adding up to the other number: for example, take 220 and 284. Proper divisors of 220 are 1, 2, 4, 5, 10, 11, 20, 22, 44, 55, and 110. They add up to 284. Proper divisors of 284 are 1, 2, 4, 71, and 142. They add up to 220. Hence, 220 and 284 are friendly!

Ancient mathematicians used friendly numbers to describe friendships between people (Aristotle in his *Ethics*, for example). And the pair of friendly numbers (also sometimes called amicable) 220 and 284 became a symbol for friendship.

An Arabic mathematician, Thabit ibn Qurra Ibrahim, who lived in Baghdad in the 9th century, came across an algorithm by which he was able to find more amicable pairs. The algorithm works in the following way:

Write the powers of 2 to the power of n in the first row, starting with $n = 1$.

In the second row write the triple of the numbers of the first row.

2	4	8	16	32
6	12	24	48	96

Add another row – in it write the number from the second row minus 1, like this

2	4	8	16	32
6	12	24	48	96
5	11	23	47	95

Add yet another row and write the product of the numbers in the second row of the column you are in and the left neighbour of this number.

2	4	8	16	32
6	12	24	48	96
5	11	23	47	95
	$12 \cdot 6 - 1 = 71$	$24 \cdot 12 - 1 = 287$	$48 \cdot 24 - 1 = 1151$	$96 \cdot 48 - 1 = 4607$

Now you need to look and try to find neighbouring primes in the third row.

In our little table that would be 5 and 11. If you had a larger table (made using exactly the same principles but for larger powers of 2), you would follow the same algorithm.

Now look at the larger of those two numbers, which is 11. Corresponding to 11 is the computed number under it, 71.

71 is in the column of 2^2, or 4. Take that 4.

You can now make the smallest pair of friendly numbers:

$$4 \cdot 5 \cdot 11 = 220$$

and

$$4 \cdot 71 = 284$$

If you replace our numbers with some letters to give us a formula, our two friendly numbers would be

	x			
y	z			
	p			

$$A = x \cdot y \cdot z$$

and

$$B = x \cdot p$$

Pierre de Fermat and Marin Mersenne discovered, in 1636, the amicable pair

$17296 = 1623 \cdot 47$ and $18416 = 16 \cdot 1151$

and Rene Descartes found the third pair

$9363584 = 128 \cdot 191 \cdot 383$ and $9437056 = 128 \cdot 73727$

In 1747 Euler, as usual, went into a bit of an overdrive and produced more amicable pairs than anyone had done before him. He published a paper *On the amicable numbers* (*De numeris amicabilius*) adding 30 more pairs; and then in three years he had extended the list to 60 amicable pairs.

Finally let's look at 10 and work out whether it has an amicable pair. From all the possible numbers, a set of friendly ones is a very small one. Real friends are rare indeed. So most integers are solitary. It may seem that this would be an easy thing to do, but no one has yet found a number that is friendly to 10.

9.5 FURTHER INVESTIGATIONS

The chapter begins with a meandering about what makes a child become what they do when they grow up. By association game (of my own) it reminds of meandering of an angel as he flies over Berlin and recites a poem written by Peter Handke in the film by Wim Wenders and Richard Reitinger, *The Skies above Berlin, Himmel über Berlin* (often translated to English as *Wings of Desire*). The film was released in 1987. If you can get hold of it, watch at least that first scene.

If you really want to learn of them, perhaps you can go to *Millennium Prize Problems* – plenty of information on those on the internet. Here's a list of the first twin primes up to 1000.

3 and 5

5 and 7

11 and 13

17 and 19

29 and 31

41 and 43

59 and 61

71 and 73

101 and 103

107 and 109

137 and 139

149 and 151

179 and 181

191 and 193

197 and 199

227 and 229

239 and 241

269 and 271

281 and 283

311 and 313

347 and 349

419 and 421

431 and 433

461 and 463

521 and 523

569 and 571

599 and 601

617 and 619

641 and 643

659 and 661

809 and 811

821 and 823

827 and 829

857 and 859

881 and 883

CHAPTER 10

The Centre of Action

THE QUOTES FROM MEDIAEVAL to modern times on the centre and the circumference – be it of the universe, the solar system, God, one's meditative practice, or our labyrinth – will be the topics that you may want to think of as you meditate on circle and its various important points. What are such points centres of?

Marius Victorinus (precise dates unknown, flourished sometimes around 400 AD) was a philosopher, public speaker, grammarian, and teacher. He was born in Africa – unknown which part – but spent all his adult life in Rome. He was an important source for the Scholastic movement later and wrote some theological works. He is ascribed the famous quote:

> God is an infinite sphere, whose centre is everywhere and whose circumference nowhere.

This wouldn't be of too much interest to us if others didn't repeat it over the centuries to come. Blaise Pascal, for example, had a little more to say about this in his *Thoughts* (*Pensées*, 1669). He turned later to theological and philosophical work, and after his mathematical discoveries were almost all done:

> The whole visible world is only an imperceptible atom in the ample bosom of nature. No idea approaches it. We may enlarge our conceptions beyond an imaginable space; we only produce atoms in comparison with the reality of things. It is an infinite sphere, the centre of which is everywhere, the circumference nowhere. In short, it is the greatest sensible mark of the almighty power of God that imagination loses itself in that thought.

DOI: 10.1201/9781003280668-10

To go in circles may be enjoyable. The next one is about the essay Jorge Luis Borges wrote, entitled *La esfera de Pascal* (*Pascal's Sphere*) in 1951. If you have, by now, read the short story *The Library of Babel* by Borges and enjoyed it, you may want to read the *Sphere of Pascal* too: here he criticises Pascal's vision of *the* sphere. As of Pascal, he said:

> [He, Pascal] Hated the universe and yearned to adore God, but God was less real to him than the hated universe. He lamented that the firmament did not speak; he compared our lives to the shipwrecked on a desert island.

Apparently, according to Borges, in an earlier version, Pascal had started to write:

> Nature is a frightful sphere, the centre of which is everywhere, and the circumference nowhere.

When you know something, it is less frightening. As we can't measure the universe, or God, or anything similar, let's have a look at the circle itself. Circumference too is incommensurable with its diameter but we won't talk about that at the moment.

10.1 IT IS ELEMENTARY, MY DEAR MISS WATSON

One of my favourite students (she was training to be a teacher) was called Miss Watson. If I were to teach her today, I would do more of the type of mathematics I keep doing here. When I *did* actually teach her, I was so worried about what *teacher competencies* she had to have, that talk about those things always took priority. I am not sure the two areas of expertise were circles which didn't coincide, so to speak.

So here's to Miss Watson. A mixture of general terms, and some thinking about what circle, really is, all about.

Circle is an Euclidean construction – you need a length of some magnitude, and a compass to describe it around the centre. The third axiom, postulate, of Euclid's *Elements* states that it is "possible to describe a circle with any centre and radius".

The following are a bit of a rambling of an old rambler, but you may want to think through this and ramble or doodle yourself.

Start from a finite length. This will be the base of your triangle.

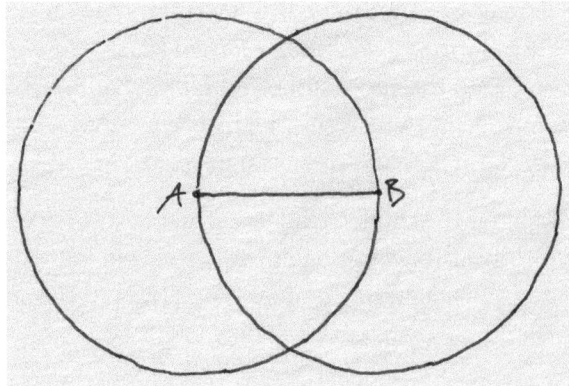

FIGURE 88   The first step of deciding on the size of an equilateral triangle – start with a base.

From A and B draw circles with radius AB. The intersection of these two circles will give two possible vertices for an equilateral triangle. Or two equilateral triangles.

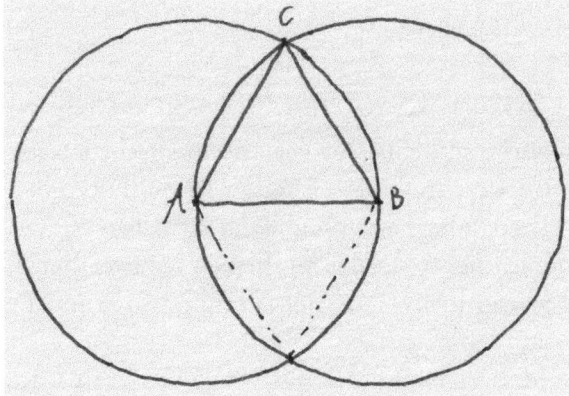

FIGURE 89   Constructing an (or two) equilateral triangle(s).

Now we know how to draw an equilateral triangle in a circle, what about if we draw some angles in it?

One of the oldest theorems (proven statements) in geometry states that the angle formed at the centre of a circle by lines originating from two points on that circle's circumference is double of the angle formed on the circumference of the circle by lines originating from the same points, as in the following diagram.

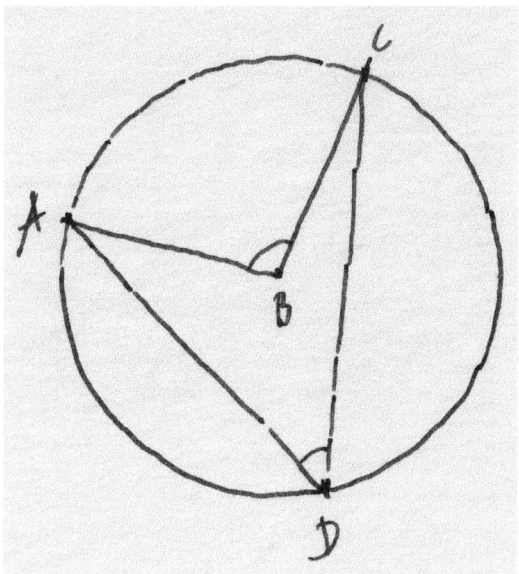

FIGURE 90   Angle at the centre theorem, $<ABC = 2 < ADC$.

This means that if A and C are on opposing sides of a diameter of the circle, the angle ADC will be equal to 90° (because $ABC = 180°$).

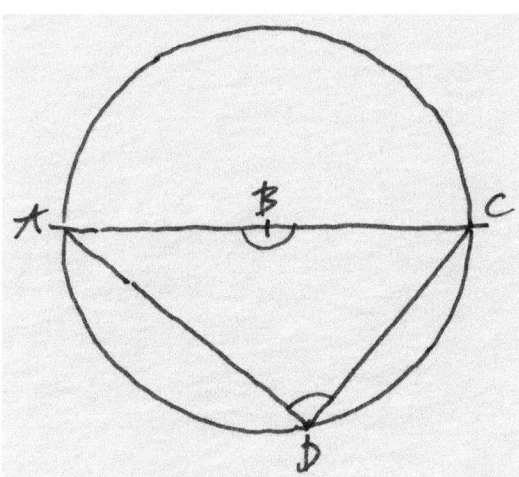

FIGURE 91   A special case of an angle at a centre.

Which leads us to conclude that a triangle in a semi-circle is a right-angled triangle. This can be very useful in some constructions which we will do later.

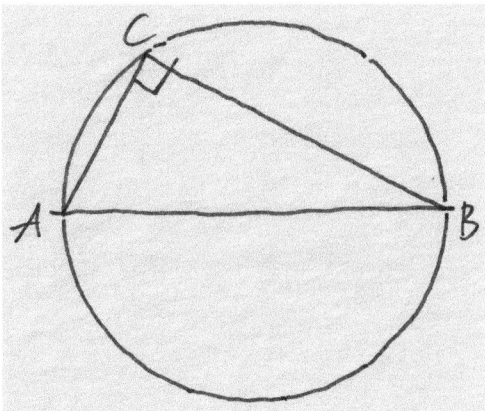

FIGURE 92   Angle in a semi-circle.

We can now use this knowledge to do a few other things.

Hexagon got its name from the Greeks – *hex* means six and this is a polygon with six edges and six vertices. The simplest construction of a hexagon is pretty simple. Start from a circle *c1*, and a point *A* on the circle. From *A* draw two circles with the same radius as *c1* (same opening of the compasses). You will end up with a pattern as follows: a construction of a hexagon through interlocking circles of the same size.

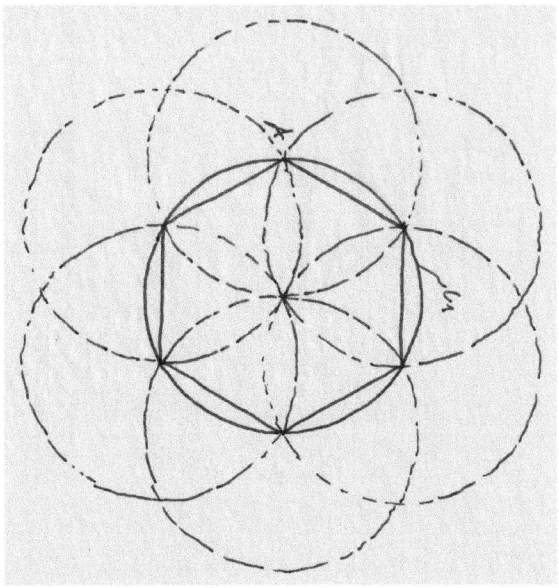

FIGURE 93   Constructing a hexagon.

Now you can just join the points of intersection between the main circle (*c1*) and the other six circles. This will give you a hexagon.

The constructions of hexagon and enneagon are related. Enneagon and nonagon mean the same thing – a nine-sided polygon. You may want to think about why this is so – and when you have done the following construction your understanding may be greater. Or just go straight to the doodle constructions and then work things out afterwards. To do that, we'll start with a circle and the hexagon inscribed in it.

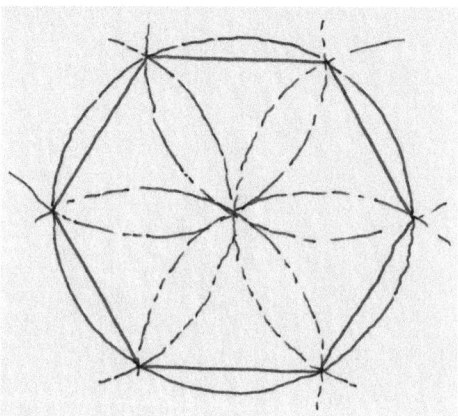

FIGURE 94   Hexagon doodle.

Inside this hexagon, you can distinguish two intertwined equilateral triangles, often called *David's star*. Make this visible.

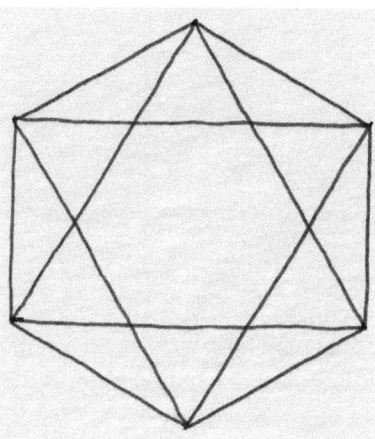

FIGURE 95   Hexagon with a rectangle stretching inside it, and David's star emerging.

It is now easier to concentrate only on the intertwined triangles and the intersections between their axes of symmetry. So let's look at these. You will notice that one of these triangles is shaded: that is because you will need to decide which one's axis of symmetry to use in your construction further. And we have decided to use the shaded triangle as the main one in our construction.

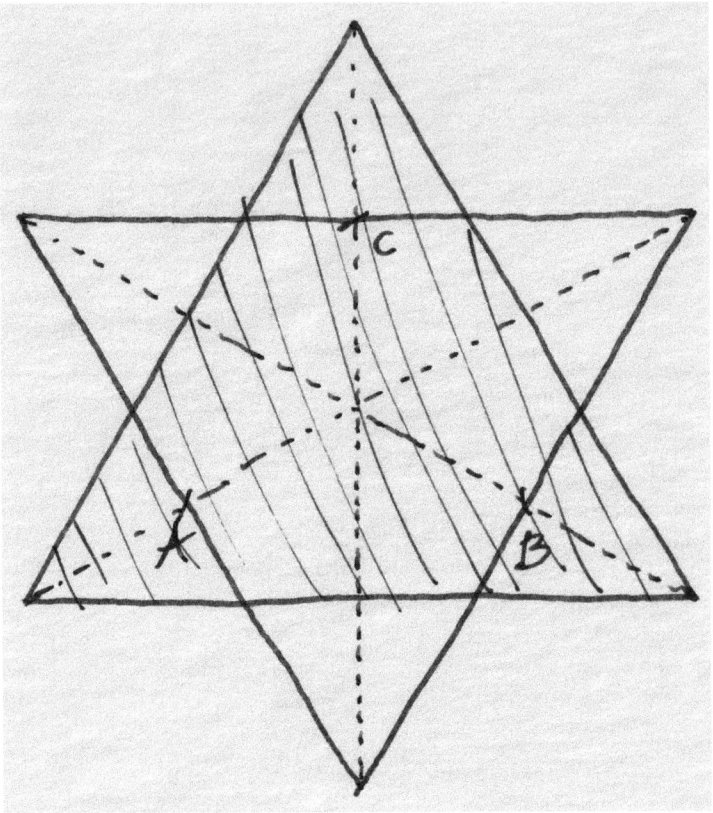

FIGURE 96   Hexagon to nonagon construction – a moment of decision through shading.

The points of intersections of the axis of symmetry of one triangle, with the sides of the other triangle are labelled as *A*, *B*, and *C*.

Use these points now to construct three arcs: from *A* as centre go through points 1 and 2; from *B* as centre, go through points 3 and 4; from *C* as centre, go through the points 5 and 6.

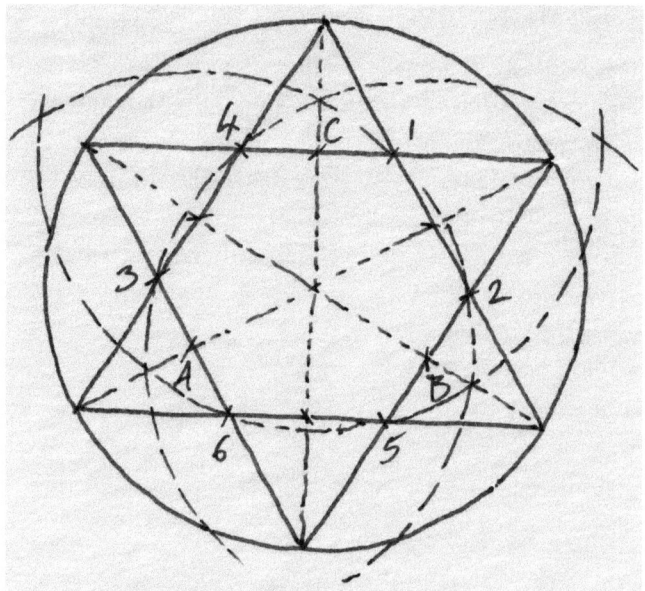

FIGURE 97   Constructing the nonagon.

These three arcs cut the original circle in six points; add the three points of the *ABC* triangle as in the previous diagram, and the circle is divided into nine sections. This gives you a nonagon = enneagon.

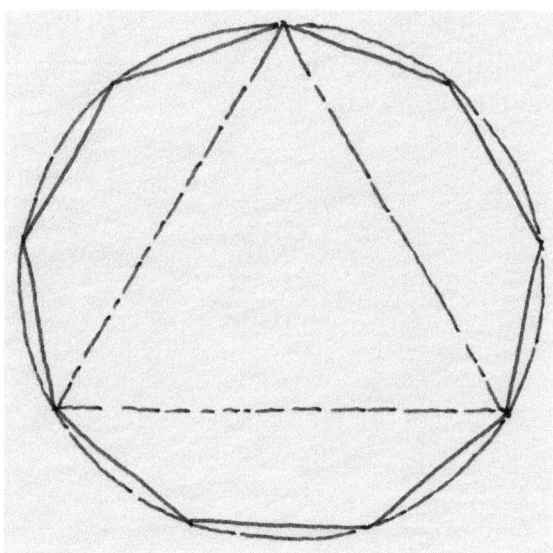

FIGURE 98   A nonagon.

10.2 A FEW OTHER CIRCLES

A circle inscribed in a triangle is called *incircle*. The centre of this circle is an intersection of all points which are equally distant from the sides of the triangle. So we need to find two (or three) lines that are *locus* of points which are all equally distant from the sides of the triangle. Locus is the set of all points that satisfy some condition. It has nothing to do with a locust. The angle bisector is such a line – it is equally distant from each point on the *arms* of the angle it bisects.

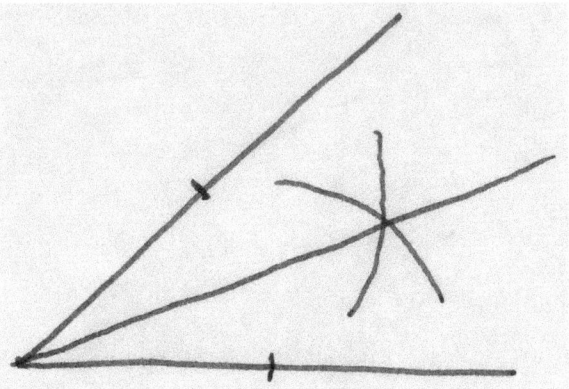

FIGURE 99   An angle bisector.

What we need to find, then, are the bisectors of all our angles (although two would suffice) of a given triangle. This intersection (of the angle bisectors) will give us the centre of the incircle.

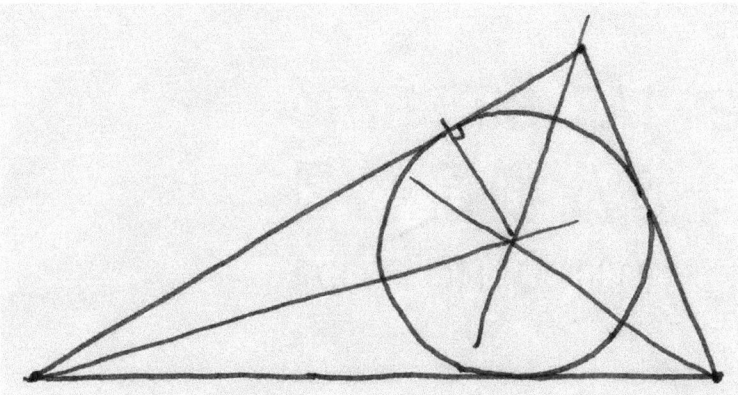

FIGURE 100   The incircle of a triangle.

The important thing to remember is that once you have the centre, you still need to find a perpendicular from this centre to one of the sides of the triangle to get a radius. A nice application of this construction would be the construction of incircles in internal triangles of various regular polygons as follows. The two given here are the incircles of the triangles that divide an equilateral triangle into three internal triangles. Try to work out which are these three internal triangles and then doodle some more. If you always divide the same equilateral triangle in different ways, and then inscribe circles in them, will the sum of areas of these circles always be the same? What about their circumferences?

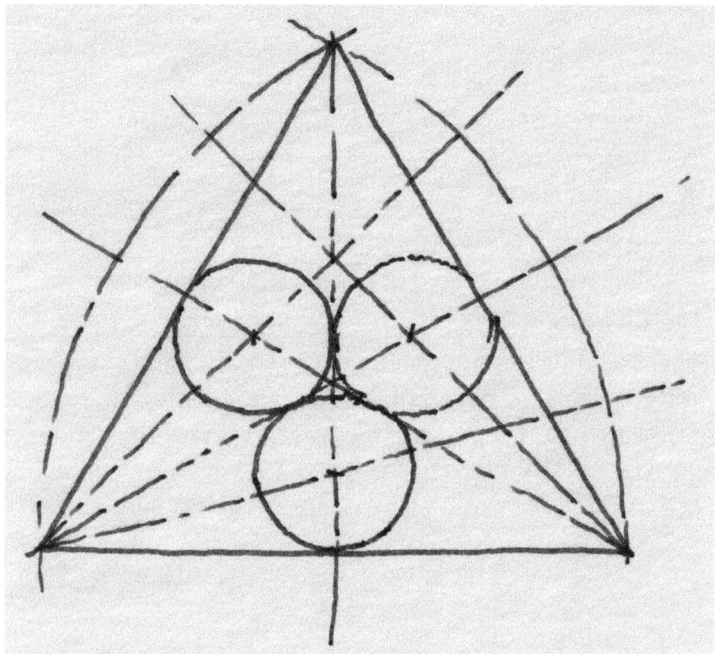

FIGURE 101   Circles inscribed in an equilateral circle.

We saw something similar with the sangaku earlier. You may want to play with more complex shapes and divide them into shapes in which you could draw incircles. Thinking about this, the areas – their sums and the ratios of their radii may be a good way to get some more mathematical joy. Here's a octagon where the radius of the middle circle is equal to the diameter of the small circles. In this octagon, inscribed in a circle, there are nine circles inscribed in it, eight of which are touching midpoints of its sides, and the largest of them is based in the centre of the octagon.

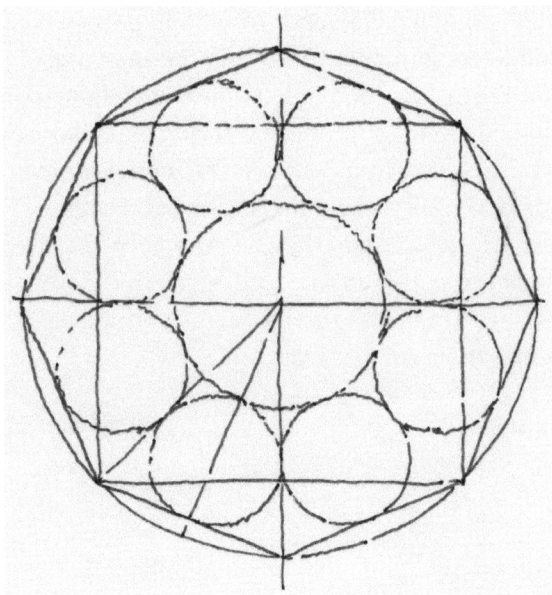

FIGURE 102   An inscribed circle collection.

10.3 THE EXCIRCLES

An excircle is a circle that is contained by two extended sides of a triangle and is on the outside of that triangle. Each triangle has three excircles (as it has three vertices). It's a fun game to see a few things about excircles in general. Start first by looking at the space where an excircle will live. Extend the two sides of a triangle.

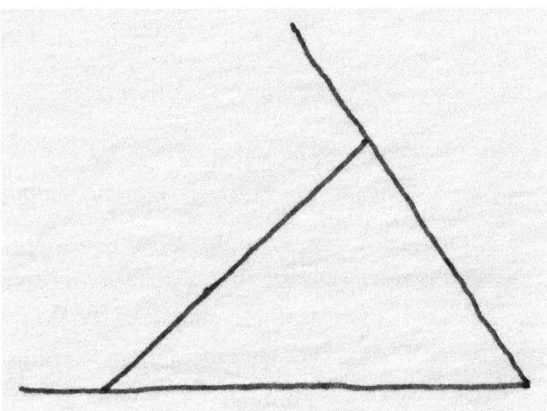

FIGURE 103   Looking at the place where our excircle will live.

This circle has to touch the third base, as well as the two extended sides. This means that the circle's centre will be equidistant from all three sides, and outside of the triangle. Construct therefore the angle bisectors of the three angles, and you will be able to find the centre of this excircle. The radius of the excircle will be found (or rather its length and where it touches one of the sides) by *dropping* a perpendicular to one of the sides. Let's do it for the side BD.

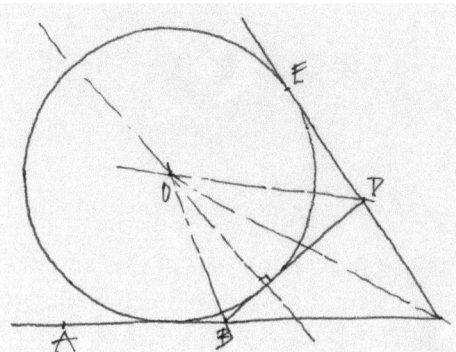

FIGURE 104   An excircle construction.

Now what is really interesting with excircles is that once they are constructed on all three sides, they will have a certain a relationship with each other, and with the incircle of the triangle. I've re-doodled the next triangle a bit, so that you could see how that works for the three excircles and the incircle.

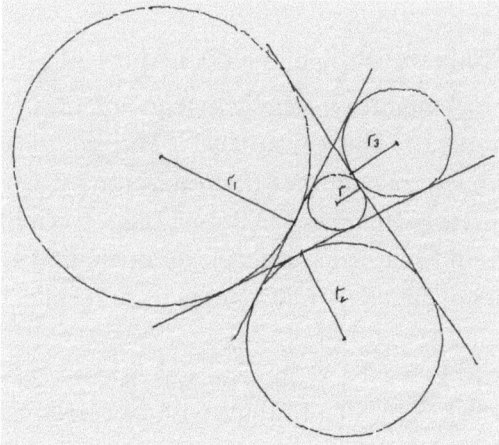

FIGURE 105   All three excircles and the incircle of a triangle.

Perhaps this is a good moment for a little pause. Think about what possible relationship these three excircles could have with the incircle. You can have some time for a pause. And the book can reserve a space for this.

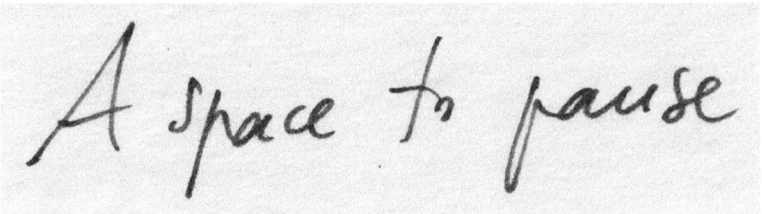

FIGURE 106   Space reserved for a moment of your thinking time.

Turns out that the sum of reciprocals of the radii of three excircles will be equal to the reciprocal of the radius of the incircle. In other words (or in mathematical symbols) that would mean

$$\frac{1}{r_1} + \frac{1}{r_2} + \frac{1}{r_3} = \frac{1}{r}$$

Considering that the circumference is of the same order (it is also a length), what would the ratios be for them? You can think, I'm sure, of other things to investigate here. But let us now turn to some other interesting things about circles.

10.4 SOME POINTS ON (ALL) CIRCLES

There are some nine points on a circle that are interesting for various reasons. Once you identify them, you have a *nine-point circle*! This is also called Euler's circle or sometimes (more recently) it is given a name of other mathematicians. This is one of those things; you will find that sometimes theorems and mathematical objects (including constructions, or special properties of circles like the ones here, or formulae) change names as times go.

We'll stick with Euler for a moment. Euler's circle passes through the intersection of the heights of a triangle with its corresponding sides (also called perpendicular feet).

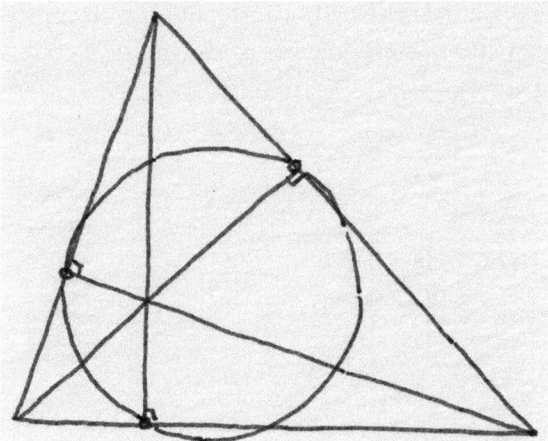

FIGURE 107   Three out of nine points – the perpendicular feet of the heights of a triangle.

These are, therefore, the three first points through which we know this circle will pass: the *feet of the heights* of the triangle. In 1765 Euler showed that this circle also passes through the midpoints of the sides of a triangle.

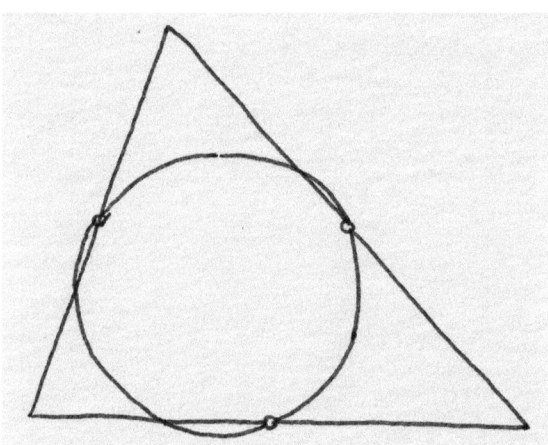

FIGURE 108   Three more points of the nine-point circle – these ones are midpoints of the circle's edges.

In 1822 Feuerbach showed that the nine-point circle passes also through the three midpoints of *segments* that join *orthocentre* with the vertices. Orthocentre is a point where all altitudes (heights) of the triangle intersect.

In an equilateral triangle the orthocentre and the circumcentre (centre of the circumcircle) and incentre (centre of the incircle) coincide.

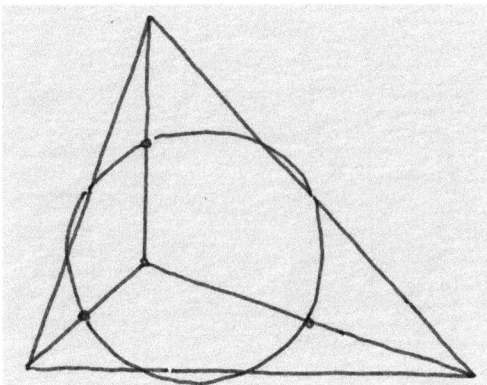

FIGURE 109   Three midpoints of the segments that join vertices with orthocentre.

So here we are, we have the nine-point circle. And the last three points are the reason the circle is also sometimes called the Feuerbach circle (Euler would have probably called it a six-point one!).

A surprising little fact is that the radius of the nine-point circle is exactly half of the radius of the circumcircle.

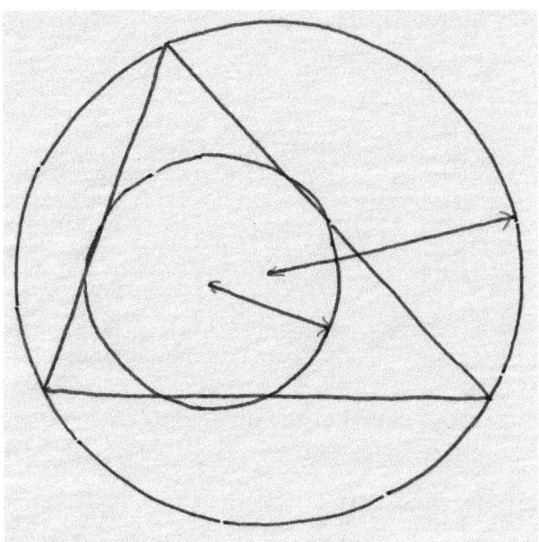

FIGURE 110   Radius of the nine-point circle is exactly half of the radius of circumcircle.

Anything else? Plenty – but not to overwhelm the meditation, let's just connect it to our previous topic – the nine-point circle is also tangent to the incircle and excircles.

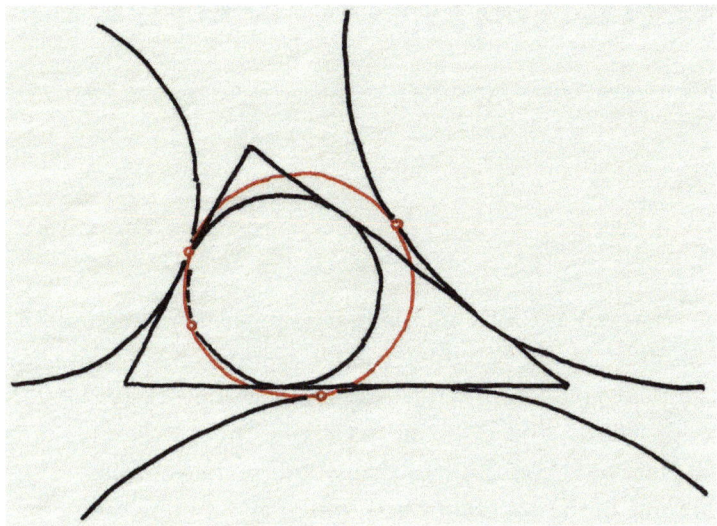

FIGURE 111   Nine-point circle is tangent to excircles and incircle.

10.5 FURTHER EXPLORATIONS

An old book by Daniel Pedoe, *Circles, a Mathematical View,* published by Dover Publications in 1979, is still a gem of a book. Try it. Not too hard, and rather beautiful.

There's a website called *Wolfram MathWorld* from the makers of *Mathematica* and *Wolfram/Alpha* which has articles about various things mentioned here and in much more detail for the dedicated. This website as well as the various other websites (and the software *Mathematica*) are highly recommended for the committed learner.

CHAPTER 11

Mathematical Objects

I F YOU WERE TO ask for a mathematical present, what would it be? For me I think it would be *Géométrie Descriptive* of Monge (I have a few editions already, but another one would always be welcome so feel free to send a copy – 18th- and 19th-century copies more welcome than others). But what else could you think of that would make your heart skip a beat? Here are some of the possibilities. Thinking about these should be as good as having them. And seeing them in person in a library or a museum even better!

11.1 *ARITHMETICA* BY DIOPHANTUS

Diophantus lived in 3rd century AD in Alexandria, Egypt. This was the centre of learning at the time ever since Alexander the Great establish the city and the library was subsequently built there. By the time of Diophantus, many a famous mathematician had passed through its streets and worked within its walls. One of them was Diophantus, of whose life we know almost nothing. However, we do know that he was born around AD 200. Michael Psellos (c. 1018–c. 1096), a Byzantine philosopher and historian, also interested in music, mentioned Diophantus in his work as someone who was versed in Egyptian arithmetic. A famous historian of mathematics, Paul Tannery (1843–1904), a brother of famous mathematician Jules Tannery (1848–1910), thought that Hypatia was the first person to describe and comment on Diophantus and his mathematics. Michael Psellos, Tannery thought, quoted from her work, alas this hasn't survived.

 DOI: 10.1201/9781003280668-11

Hypatia (c. 370–c. 415) was a daughter of mathematician Theon of Alexandria and became a famous mathematician and philosopher herself. She is often mentioned as the first known woman mathematician. That has been recently challenged by research that has uncovered names of some Pythagorean women, such as Theano but also Timycha, Philtys, Occelo and Eccelo, Cratesicleia.

Arithmetica was written sometime in the 3rd century and was a collection of 130 problems. We now call such types of problems *Diophantine equations*, and the method after the author of this book *Diophantine analysis*. What kind of equation would this be?

It would a polynomial in two or more unknowns with integer coefficients. Mathematics can sound like a gobbledegook to those who don't speak it, but it has developed in a way to make things easier to mathematicians (and probably not so for others). Translated polynomial (into a mathematical language) is an expression such as

$$x^2 + y^2 = z^2$$

Unknown quantities here are x, y, and z. And the coefficients in this particular equation are all one – just we don't need to write them. The coefficient is the value that multiplies the unknown quantity. If I wanted to write all the coefficients in the previous equation, it would look like this

$$1x^2 + 1y^2 = 1z^2$$

Which would look a bit silly and that's why I didn't do that in the first place. Diophantus just liked coefficients which were whole numbers – integers.

Diophantus would not have written this using the same notation, mind you – that was all invented centuries after him. He would have described his polynomials in a similar manner as Metrodorus did (see 13.4). Diophantus was satisfied only with rational solutions, but not with the negative solutions, as negative numbers weren't considered *proper* in his time. His book wasn't fully appreciated until the number enthusiasts of the 17th century got hold of it.

Claude Bachet (1581–1638) published a Latin translation of books I–VI in 1621. This became a very popular book in French mathematical circles. The book had many editions afterwards, and one of them was published later on by the son of Pierre de Fermat (1607–1665) in 1670, Samuel Fermat. In this one, Samuel included the notes his father left in the margin of the edition he was reading. In fact he was reading precisely on the equation we gave earlier –

$$x^2 + y^2 = z^2$$

and was asking himself whether there are solutions to this equation for any powers larger than two. In other words, would for any values of x, y, and z be possible to have this equation hold for powers that are larger than two?

Fermat (the father) wrote next to it:

I have discovered a truly remarkable proof which this margin is too small to contain.

The 1670 edition by his son was updated to include this sentence, and that became known as Fermat's Last Theorem (as it was found only after his death). It was proven only in the 20th century that Fermat was right.

But what happened to the original book Fermat was reading and in which he wrote in the margin? Librarians of the world unite and try to find it! That would be a nice one to see and hold.

11.2 HILBERT'S HOTEL

David Hilbert had an idea: if you have a hotel with a limited number of rooms there is always a chance you will be full and not be able to take any further guests. But what if you have a hotel with an infinite number of rooms? They still begin as in any ordinary hotel, from 1, 2, 3, and so on.

Sounds great? Yes, but let's imagine a new guest coming though – how can you even tell the number so far out there in the infinity? You can't check really, let alone make that room available. But if you start from those rooms that are near to you, 1, 2, 3 and so on, you can do that. Just move all the guests one step forward. Every guest goes from the room they are in to a room which is one number higher. From 1 to 2, from 2 to 3, and from some very large number a to $a + 1$.

This is all well and good and you can now have infinitely many guests. ∞ is so large we can't know it, but even the infinite set of integers is denumerable. It means they are somehow countable, but still an infinitely large set. Let's see another set like that. In our hotel instead of saying go to the room with a number $a + 1$, let's say go to a room with a number $a + 2$.

We'll get a one-to-one correspondence

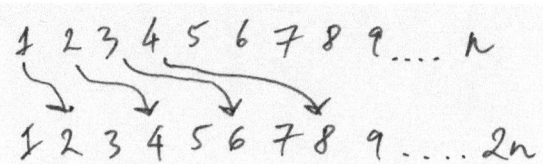

FIGURE 112   Establishing one-to-one correspondence between natural and even numbers.

You've kind of extended your infinity as there is every odd number that makes another infinity.

But you can add a third row to these infinite lists of numbers by adding a list of primes. We know that there are infinitely many primes, thanks to a proof Euclid came up with. The proof goes like this. Say you have all the primes. Then, as primes build other numbers by multiplying, they can be multiplied to come up with a number. Now let's say that number is some P, a product of all our primes. And then you can say, well to that P I will add 1, and get q. My new number will be $q = P + 1$. If this new number q is a prime itself, we got our next prime (so there is always a next one). If q is not a prime, then there will be some prime factor that divides it. But if this prime factor was in our list in the first place, it would divide P and so it couldn't divide $P + 1$ as well.

So there is always another new prime. This is why things like *Great Internet Mersenne Prime Search* exist. A never-ending story. And now we can put all our infinities under each other and see that they will be 'listable', denumerable.

Let's try to do something similar with the fractions. We can then draw a diagram showing enumeration of fractions where the top row shows fractions with 1 in numerator and natural numbers is denominator; every subsequent row having same denominators, but numerators going from 2 to 4.

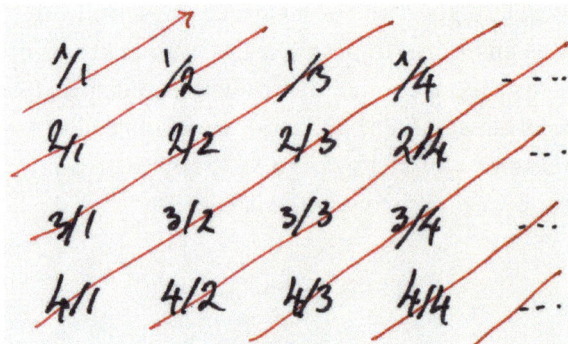

FIGURE 113   How to enumerate fractions.

So now we can connect these somehow. If we connect them diagonally, we will list all of the fractions. The job is not finished yet. We can then connect each of these with natural numbers. Because we have established such a list, and that list (natural numbers) is itself infinite, we proved that the set of rational numbers is also infinite. Rational numbers are just fractions. As long as both parts are whole numbers and the denominator is not 0.

But the only reason I've been able to do this is that I can find some order among these in terms of listing them and finding their corresponding numbers in another list. I can't actually do that with the real numbers; the infinity between 0 and 1, however small it may seem at first, is not like that and I can't list all the numbers between these two values.

We can number the hotels in our Grand Hilbert Hotel in any listable way, but not in a way that is not denumerable.

FIGURE 114   Grand Hotel Infinity.

11.3 MÖBIUS BAND

Thinking of infinity, we can go on an infinite walk around a band (not a music one) called after its inventor, Möbius. August Ferdinand Möbius (1790–1868) was a German mathematician and astronomer from Saxony. This band was independently also discovered by another German mathematician, Johann Benedict Listing (1808–1882). But we still call this Möbius band. It is a non-orientable surface. You can't distinguish clockwise with a non-clockwise direction on this surface.

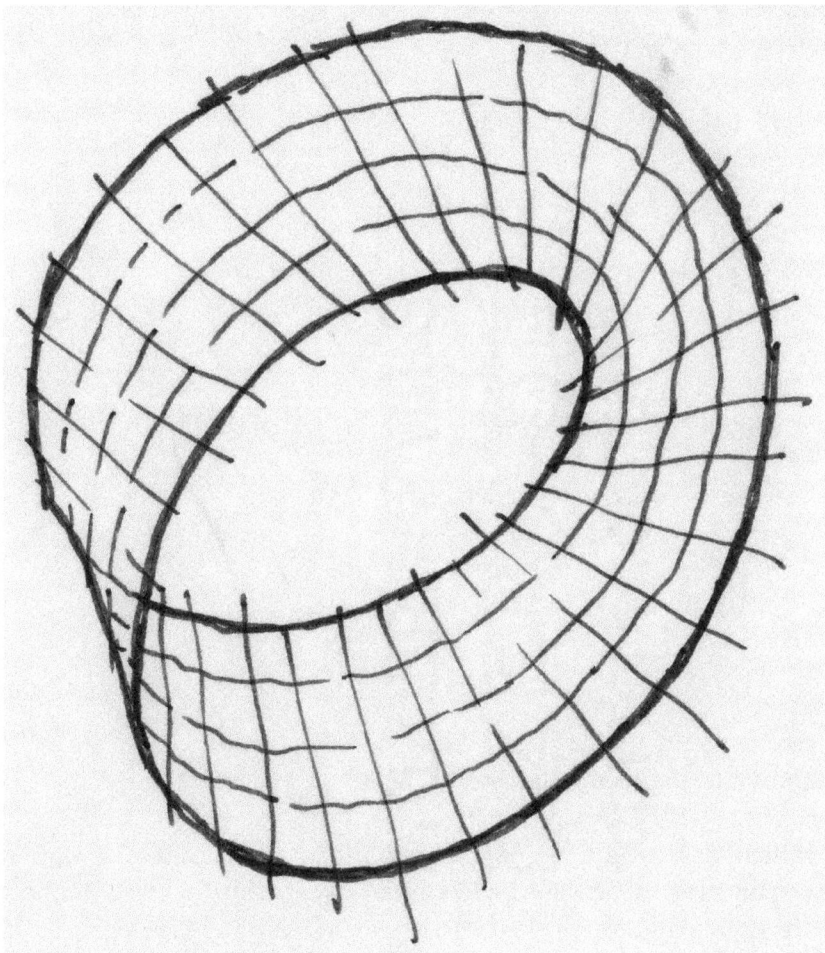

FIGURE 115   The band of Möbius.

You can start from one place and soon you will find yourself on the other side of that same shape. (Make it with a strip of paper which you twist once and fasten with a tape.)

Although Möbius described it and studied it mathematically, there are images of the same kind of surfaces in much older depictions, such as for example an ancient mosaic from a Roman villa in Sentinum, Italy, dated to about AD 200–250.

In this image Hellenic God Aion holds the strip which has signs of the zodiac on it. This is a God of time and its possible cyclic nature.

FIGURE 116   God Aion with Möbius band.

Millennia later, a Christian depiction of a similar sense of time was drawn on a fresco, half of which only survives at the Rochester Cathedral in Kent, England. Here the time is controlled by Queen Fortuna. She directs the turning of the wheel with the most fortunate person at any one time sitting on the top of the wheel but soon falling off as time progresses.

FIGURE 117   A rather poor depiction of the goddess of fortune sitting on her wheel.

Indeed, the nature of time is one of the greatest unsolved problems of mathematics, physics, and metaphysics. Does it really move forward or is this our experience only? Does everything exist all at once? Some would suggest it does.

In any case, if you wish to meditate upon the question of time, there are worse things you could do than using some of your time on creating Möbius strips. Multiple? Why not – you can make one then try other ways of doing different things with it. Twist the strip twice then tape it. Then do another, twist the strip three times, etc. Then you can also make a very wide Möbius strip and cut it in the middle and see what happens. What happens if you cut such a wide strip into three strips (once you have taped it)?

11.4 PIERO DELLA FRANCESCA'S MASTERPIECE?

Meditation has a great power to take the mind to a place of peace and beauty through recreating images from nature, or as we have been doing, mathematics. But what if you are concerned by some turmoil that is taking place? Remember the introduction to the book. Mathematics can be and is used sometimes to create an image, a model of the world that one can study and understand better. In that respect, it is a little bit like magic.

Sometimes you can connect this *modelling* with art. Take some seriously beautiful painting that portrays such a turmoil and study it using mathematics of some kind. You can use linear perspective, or any other method.

One of my favourite techniques of contemplation is recreating an art scene through mathematics. Who could be better than Piero della Francesca for such an exercise? Both a mathematician and an amazing artist, he gives us many opportunities to study his paintings through mathematics. I can't make my mind whether I would rather like to see him working on his mathematical manuscripts and recreating Archimedean solids, or doing his paintings.

Of all his paintings, his *The Flagellation of Christ*, painted sometime between 1455 and 1460, is my favourite. It shows, as its title says, the flagellation of Christ by the Romans. The biblical event, though, takes place in this picture in the background, and the Christ is bound to a pillar upon which a golden figure of God Hermes holds a small ball as if to make an offering of a kind. Hermes was a Roman god that was able to move between the worlds of ordinary mortals and the divine, and was meant to be both an inspiration and a protector of souls.

FIGURE 118   A doodle of *The Flagellation of Christ* by Francesca.

The importance of the painting for art history is in its use of linear perspective that Piero knew so well. He contributed to both the development of its theory and the masterly application of it in practice. This painting in particular is known as one of the best examples of how to construct a painting using linear perspective. The scene was therefore reconstructed twice in the 20th century by leading historians of art. It showed that all the elements of the image were absolutely perfectly depicted, bar a line that comes in the middle dividing the group of somewhat worried men in the foreground from the violence that is unleashed on the young man in the background. The buildings behind the three front figures are also slightly distorted for artistic effect.

But it also portrays, some argue without doubt, a message Cardinal Bessarion (1403–1472) wanted Piero to depict. Bessarion was a Byzantine scholar, Patriarch of Constantinople, who sought refuge in Italy and remained there to the end of his life. He translated many works, which would have otherwise surely perished, to the Latin West. The painting, it is believed, shows the fall of the Eastern Roman Empire, through the fall of Constantinople, and the beginning of the new world order as it then began through the second crusade that began soon after. The light that emanated from Hermes' hand holding a small ball has intrigued historians of art for a long time. Technically this painting is perfectly constructed. But what could it mean? The meditative scene invokes with near-perfect geometric precision an imagined scene of great significance.

This painting is housed in the Ducal Palace in Urbino with some other masterpieces of Renaissance art. Perhaps only a handful of paintings from this period would compare to the *Flagellation*, and none with greater use of geometry to create such an immensely powerful and utterly static image of the great turmoil that would sweep through Europe from then on. In fact, that is where I think its power lies, in our ability to consume such turmoil with such great exactitude, with a help of mathematics, and with the help of the great artist.

11.5 FURTHER READING

A good historical account of Pythagorean women is given in Dorota Dutsch's *Pythagorean Women Philosophers: Between Belief and Suspicion*, Oxford University Press, 2010.

You could also see the blog from the Faculty of Mathematics, Cambridge, on *Fermat's Last Theorem – From History to New Mathematics* (accessed 25 May 2024 www.maths.cam.ac.uk/features/fermats-last-theorem-history-new-mathematics).

Piero della Francesca's *Flagellation* can be seen in the Galleria Nazionale delle Marche in Urbino, in Italy. High time to go there and see it.

CHAPTER 12

Tracing the Movement

ABOUT 250 (OR SO) years ago a young student was given a boring job. He was a son of a merchant, and although he was talented, there was not much money in the family for him to plan for an academic or military career. During the summer holidays, however, he drew a map of his birthplace, a rich city of Beaune in the Burgundy region in France. The map was so fantastically good that a place was offered to him in the drafting office in the then newly founded École Royale du Génie de Mézières. So now he was in, his superior gave him a job to determine the height of a fortification which was being designed.

Until then, there were two methods used for this problem. One involved choosing the most characteristic points on the terrain surrounding the fortification, and constructing the triangles determined by the viewpoint. The point of the edge of the fortification and the height of the wall sufficient to offer effective protection were crucial in this design. The whole point of the fortification was to offer protection to those inside.

The other method of finding what should the height of the wall was based on long calculations, with the height of each crucial point being measured directly on the terrain and noted on a plan.

Gaspard Monge, that was the name of the young man, had a different idea. His plan had two initial stages. Firstly, he chose a few of the highest points from the surrounding terrain. Through these, he drew tangents to the fortification, adding sufficient height to the wall to protect the fortification from missiles. He then used these tangent lines to generate a tangential surface to the terrain. Doing this, he was able to reduce the time needed for the calculations very considerably, as all other points were under this tangential surface.

The method, when first explained to his supervisors and then understood by them, was immediately ruled a military secret. From there on, Monge perfected his method but was unable to publish anything about it until a reform of the whole educational system took place during the Revolution, and until the ban on publishing his method was lifted. Considering he was by then an influential leader of the new Republic and one of the main founders of the new schools of the Republic, the ban was lifted.

After the Revolution, descriptive geometry was taught in France and its colonies as well as in all the lands French had the cultural or educational influence on.

Most people disregard this technique as a graphical technique only. But it is a beautiful and imaginative way of imagining space that allowed Monge to develop new mathematics and connect analysis and geometry, and one that has been often missed in modern interpretations. Analysis deals with analysing functions and curves they produce in detail.

It's also a great way to let yourself explore space in the ways you probably haven't done before. We will concentrate on Monge's view of space and how it is generated.

12.1 IMAGINE A TRACE YOU LEAVE BEHIND

In geometry we tend to inherit the ways of defining things even if we don't agree with the definitions. The structure of Euclid's *Elements*, written around 300 BC, continued to rule geometric investigations for more than 20 centuries. But one geometrical technique was different although it still counts as Euclidean geometry. It was the descriptive geometry of Monge. This technique didn't take the elements of geometry to be independent from each other. Everything, in this technique, was generated from a point, and that point, when moved, did all there was to do in creating the whole space.

Let's begin by thinking about this. Imagine a point in space. A point is that which has no part, and is without width, length, or thickness. It is without dimension. You may want to look far away or close your eyes to be able to imagine such a point more easily.

Imagine now that this point starts moving in a straight line. As it moves, it leaves a trace.

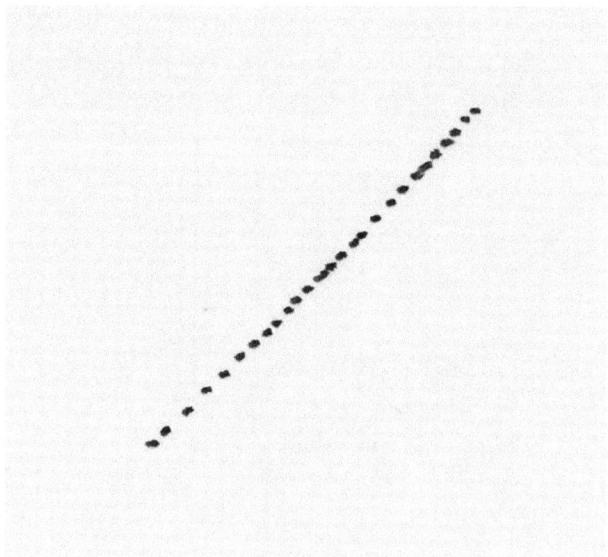

FIGURE 119   Movement of a point traces a line.

Eventually it will get into infinity, but the trace will stay behind.

Let that trace, the straight line, solidify and become our next element. It will have only one dimension, its length.

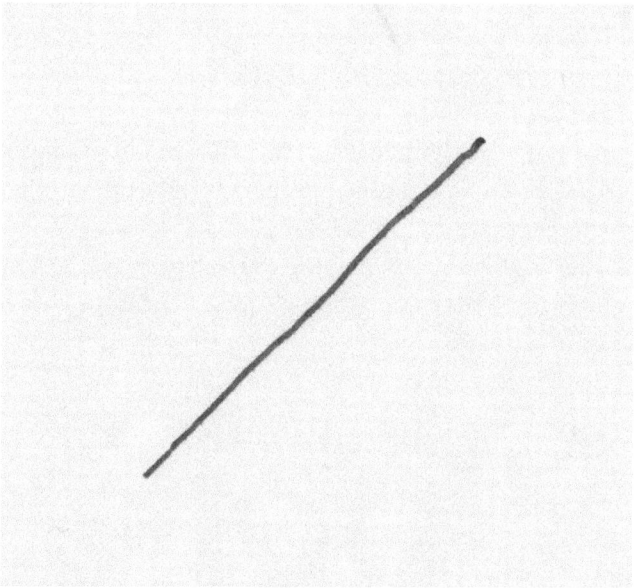

FIGURE 120   A line that is now left behind.

Now imagine if that straight line is stationary. Another straight line like it will now cross it at a right angle. Imagine now that this second line starts moving across the first one. As it does so, it too leaves a trace behind.

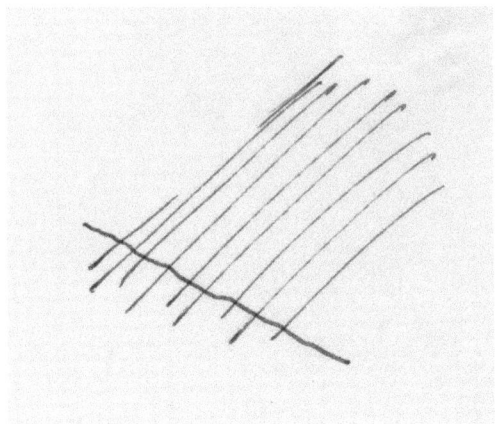

FIGURE 121　The line crossed by another line, which in turn moves along it.

When this trace of the line solidifies, it will create a flat plane. A plane will have two dimensions: length and breadth.

Imagine now that this plane becomes your next element. At a perpendicular angle let another line serve as a direction for this plane to move. As it does, it will leave a trace. We have just created a three-dimensional space.

This is the time to start imposing restrictions and make some objects. Let's do the cube.

Start from a point but let it leave a trace of a singular length. From one endpoint, at a right angle imagine another straight line of the exact same length, move to generate a square.

At one corner of this square imagine the line of the same length as original, at a right angle to both edges of the square. You have now constructed a cube in your imagination.

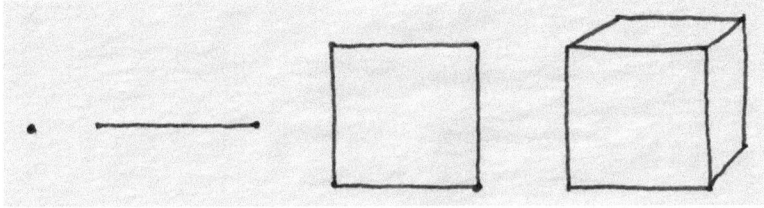

FIGURE 122　From point to cube.

Imagine now that this cube is in a space of *n* dimensions. We can't necessarily visualise it, but it can be constructed just in the same way. Use the segment of the same length as original and set it orthogonally to the three-dimensional cube and move the cube in that direction. You would have created a hypercube. There are various depictions of it in mathematics textbooks and in animations. Try to find some animations of it. Can you visualise the fourth dimension yourself? You can try to play with that as much as you like.

Alicia Boole Stott (1860–1940) was able to see things in the fourth dimension. She constructed objects in third dimension which are projections of the objects from the fourth dimension. Here is one of the objects – a fourth dimensional object drawn here in the second dimension.

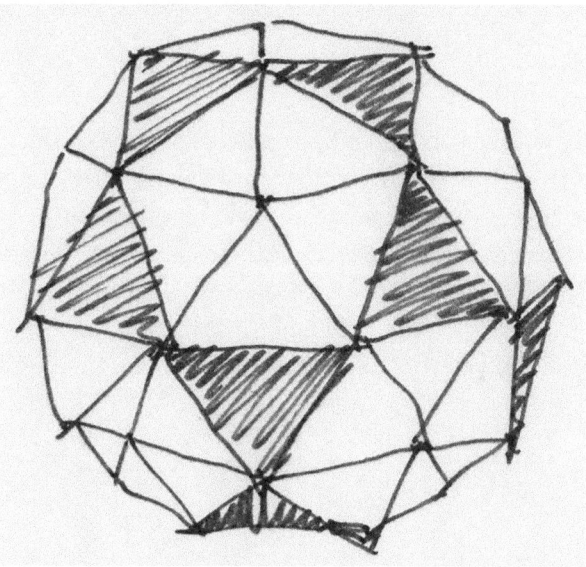

FIGURE 123   600-cell's third-dimensional projection, then projected onto a two-dimensional drawing.

This is a projection of the central part of the 600-cell, which is a convex regular four-dimensional polytope. Like Platonic solids are regular polytopes in three dimensions, 600-cell is a regular polytope in the fourth dimension. It has 600 tetrahedral cells, with 20 of them meeting at each vertex. You can search for some other regular polytopes from the fourth dimension. Although they live in that dimension, once they pass through our three-dimensional space they will leave some traces, similar to the objects like this one that Alicia made.

12.2 HYPERBOLAS AND ELLIPSES IN MOTION

There are beautiful things you can construct in your meditations by thinking about very simple objects. Imagine a circle in a plane.

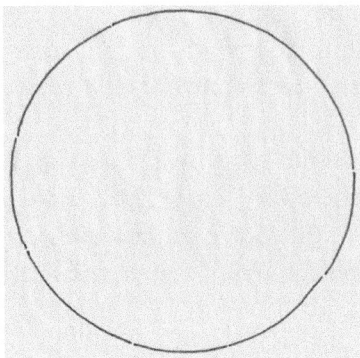

FIGURE 124   A circle in a plane.

Imagine now that this circle has copied itself. Now imagine the copy to lift above this circle (into a third dimension). Now you can imagine that you have two circles one exactly above the other in three-dimensional space. From this circle, now *drop* a segment (a line) to join a point on the bottom circle. Now move one circle around for a bit (say 1/4 of a turn). Your line, joining two circles, will have to move a bit too. But how?

From the side, if you look at it, what we just imagined will look something like this.

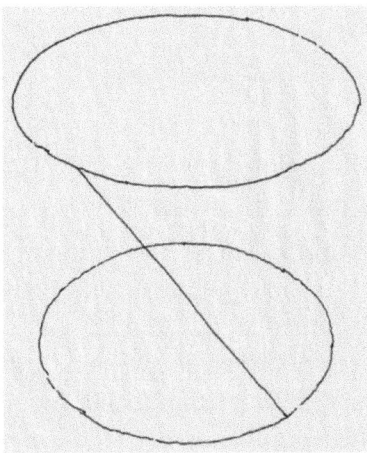

FIGURE 125   A hyperboloid, just about to be generated.

Not a huge deal so far. Now imagine that this line that connects two circles and is at an angle to both of them starts moving around on these two circles. Trace of this segment will generate a surface of a hyperboloid.

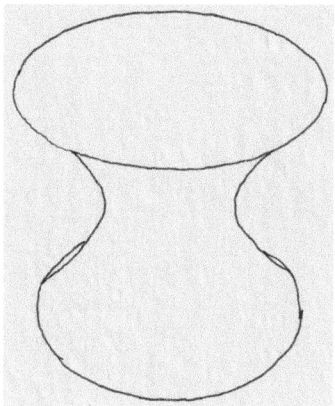

FIGURE 126   Hyperboloid.

If you cut this object with a flat plane, perpendicular at the circles that we used to make it, you will get hyperbolas. A hyperboloid can also be generated by rotating a hyperbola around one of its principal axes. So you could have just imagined a hyperbola and then rotated one of its 'sheets' around the y axis to create the same kind of object.

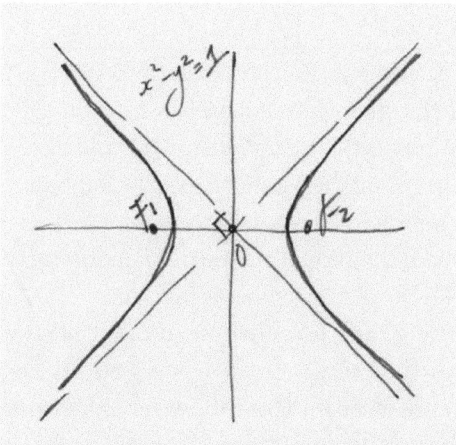

FIGURE 127   Hyperbola, which you could also spin around its axis to create a hyperboloid.

Imagine now a square in a flat surface. From this square pick two opposite vertices and construct perpendiculars to them. Now imagine lifting the square's two vertices to these new points and leaving the others in the original plane. You will get a hyperboloid paraboloid. This shape was quite popular in architectural design in the middle (or a little after) of the 20th century. For example, a railway station in Warsaw has a surface like that for its roof (built in 1962).

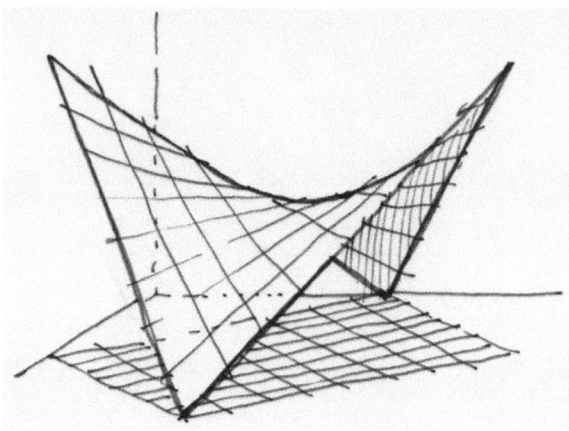

FIGURE 128   Hyperboloid paraboloid.

There is plenty to be imagined through Monge's principle of leaving traces and playing in your mind with curves of points moving and generating new objects.

12.3 DANDELIN'S ICE CREAM

Imagining space became easy for students of École Polytechnique of which Monge was one of the main founders. He was even called the father of the school. One of the students there was Germinal Dandelin (1794–1847). His first name comes from the month he was born in – revolutionary month of *Germinal* which was from 21 March to 20 April. He came up with some other mathematical inventions, but we'll mention him here for something that is now known as *Dandelin spheres*. He showed that spheres in a cone act in certain ways.

Before getting there, let's establish what conic sections are there. There are different things that can happen when you take a flat plane and cut a cone (double cone) with it. The following illustration shows what the options are. The three bottom pictures are *degenerate* sections, which usually don't count. The ones that are interesting are the four top ones. They depend on the angle at which this plane cuts the cone – and through that creates parabola, ellipse, circle, and hyperbola.

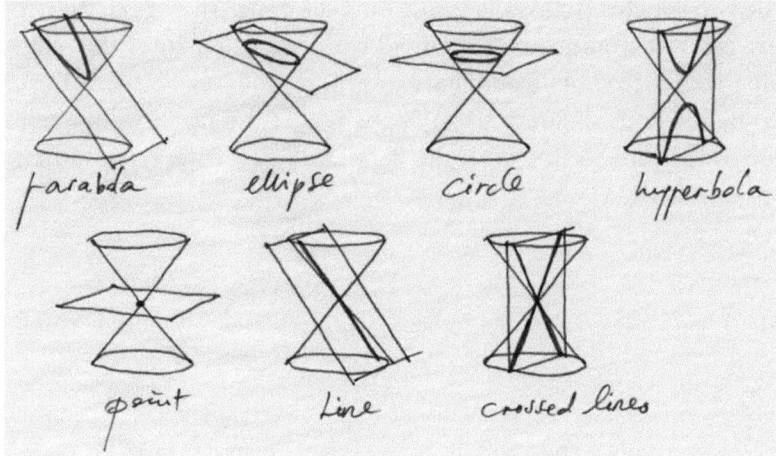

FIGURE 129   Conic sections.

So let's imagine a cone with some ice cream at the top and one lower down within the cone. The scoops are perfect, and definitely not runny!

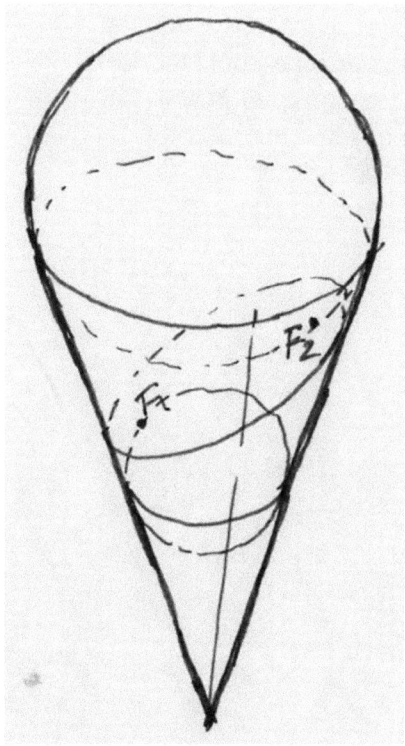

FIGURE 130   Dandelin ice cream set-up.

The two spheres (scoops) are touching the cone. That's pretty obvious as otherwise they wouldn't be contained within it. The part of the plane that would touch them both would have a curious property. You can already see it on the previous diagram. It so happens that the plane would create an ellipse in the intersection with the cone. And that ellipse would have two foci as any other.

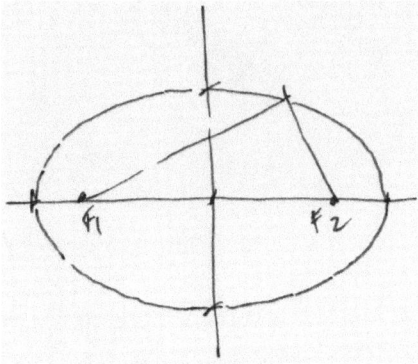

FIGURE 131   Ellipse.

The two spheres of our cone would touch the ellipse in its *foci*! Isn't that just extraordinary? The question now is: what happens if you have only one sphere or scoop in your cone, and you cut the cone in a different way?

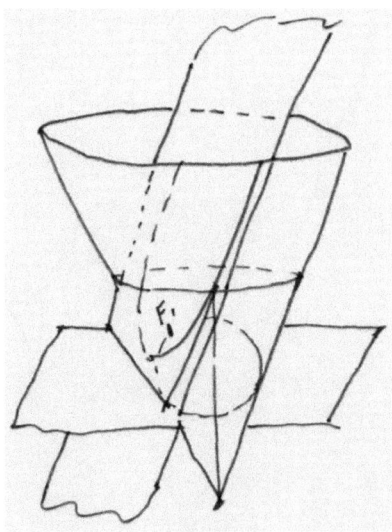

FIGURE 132   A cone with a rather small sphere in it.

This time we have one sphere, and a plane at a different angle. The angle this plane goes to cut the cone means that it will produce a parabola; and as in the previous case, the parabola will touch the sphere in its focus.

If you wanted to see what happens with hyperbola, you will need to extend your cone into the other direction too.

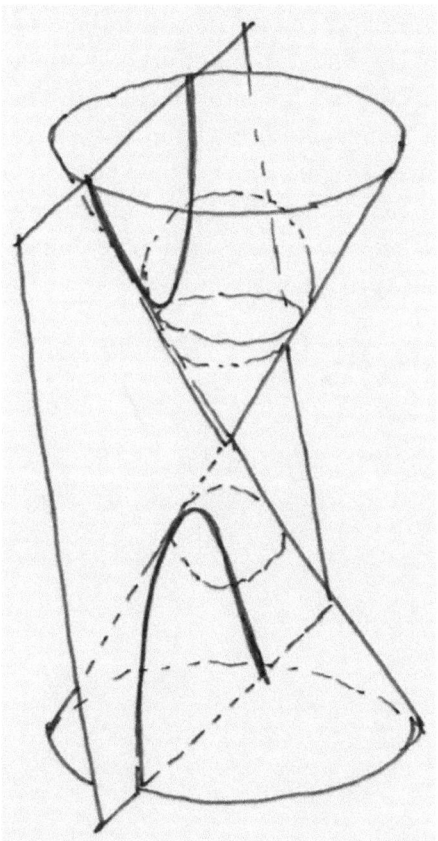

FIGURE 133   Dandelin spheres touch hyperbolae in its foci.

Depending on how you position your spheres (or the intersecting plane), you should have two spheres of equal size. It doesn't matter if the spheres are not the same size, though, as the plane that cuts both parts of the double cone. It will touch two spheres in the foci of the hyperbola. Hyperbola is the curve you will get as an intersection of the plane and the double cone.

12.4 FURTHER EXPLORATIONS

In the introduction to the chapter 11, I used a phrase about something making *your heart skip a beat*. In French, that is *le coup de cour*. By association, there's a film noir that is one of my favourites – *The Beat That My Heart Skipped*, which is kind of a similar phrase, but the original French title is *De battre mon cœur s'est arrêté*. It was released in 2005. There's a fair amount of violence there so I wouldn't recommend it to the faint-hearted, but the lesson, the message, and the nature of the *beat* that the main character's heart skipped is well worth the effort of finding this film and investing a couple of hours of your life into watching it. I have no idea why this film reminds me of Monge. But Monge too was a boy who fought through quite a few problematic moments in his life and through some tricky situations. Not least the Revolution, the Terror, and then towards the end, the fall of his friend, Napoleon. He managed to get out of it all, and stand up tall on the other side, through mathematics.

CHAPTER 13

Mathematicians

FIGURE 134   Alicia Boole Stott.

13.1 ALICIA BOOLE STOTT (1860–1940)

Alicia was a British mathematician, a daughter of George and Mary Boole. From George (1815–1864) we have Boolean algebra which is based on binary system of 0s and 1s. And from this it was possible, a century after he invented it, to get computers to work in the way they work today.

DOI: 10.1201/9781003280668-13

As a child Alicia was taught how to "see" things in fourth dimension by her sister's boyfriend, later husband, Charles Hinton. Alicia was also taught mathematics by her mother but later was largely self-taught. She ended up having a doctorate in mathematics bestowed upon her by the University of Groningen through her collaboration with their professor of mathematics Pieter Hendrik Schoute (1846–1913). She became an influential figure in mathematics in her later life as she made all these beautiful mathematical models of four-dimensional objects passing through three dimensions.

The whole of the Boole family are a fascinating lot, and you could do worse than investigate their many members through some further reading. Highly recommended comes the novel by Alicia's sister, Ethel Lilian Voynich (1864–1960), *The Gadfly*, which inspired many a revolutionary in Russia and even China. She was in turn married to Michal Habdank-Wojnicz (1865–1930), a Polish revolutionary from whom there is still a Voynich manuscript left for the cryptographers to decipher.

FIGURE 135   Viggo Brun.

13.2 VIGGO BRUN (1885–1978)

He was a Norwegian mathematician interested in numbers. He introduced a sieve of prime numbers that was different to the ancient one of Eratosthenes (II cent BC), yet one that gave pretty much the same results. This was about finding prime numbers by *sieving* them. We mention Viggo because he was interested in prime numbers and especially the twin primes.

There's a theorem called Brun's theorem. It says that the sum of the reciprocals of twin primes converges to a finite value known as Brun's constant. This constant itself is a number which is a limit of the sum of reciprocals of odd twin primes:

$$B \equiv \left(\frac{1}{2}+\frac{1}{3}\right)+\left(\frac{1}{5}+\frac{1}{7}\right)+\left(\frac{1}{11}+\frac{1}{13}\right)+\left(\frac{1}{17}+\frac{1}{19}\right)\dots$$

FIGURE 136   Germinal Dandelin.

13.3 GERMINAL DANDELIN (1794–1847)

Germinal was a French mathematician who we have mentioned here because of his works on what is now known as Dandelin's spheres. A European by birth as well as in heart, he was born near Paris (his mother was Belgian), studied in Ghent and then at the École Polytechnique, and fought in Napoleon's army. He later became a citizen of Netherlands and a professor of mining engineering in Belgium.

FIGURE 137   Diophantus.

13.4 DIOPHANTUS (DATE OF BIRTH UNKNOWN? – 284 AD)

Diophantus was a Greek mathematician whose age we should all be able to find through a puzzle left behind by Metrodorus (dates unknown, around 6th century AD), another Greek. According to Metrodorus, Diophantus' boyhood was one-sixth of his life, one-twelfth more, then after one-seventh he got married. In five years he had a son, but he died after he was one-half of his father's age. After four years Diophantus himself died.

But apart from this puzzle, we have polygonal numbers that he studied, equations that are called after him, and adoration from most number theorists the world over.

Freeman Dyson 1923-2020

FIGURE 138   Dyson.

13.5 FREEMAN DYSON (1923–2020)

Dyson was a British-American mathematician and physicist. He became a professor at the Princeton Institute of Advanced Studies and came up with many new concepts in applied mathematics and physics.

In this book Dyson appears in the chapter on Japanese sangaku, but I became interested in him once I read his book *From Eros to Gaia*. This is a wonderful little book, published by Pantheon Books in 1992. It is a non-fiction book, a bit like this one, with a lot of personal experiences and thoughts thrown in. Highly recommended.

FIGURE 139   Euclid.

13.6 EUCLID

Euclid (not of Megara, but of Alexandria), lived around 300 BC when his *Elements* were published. He was a geometrician and logician and is often called the father of geometry. In mediaeval Europe, during the Middle Ages, Euclid appeared in the Old Masonic Charges, which were a kind of very archaic history of mathematics written in mediaeval Europe for masons travelling around and building magnificent cathedrals.

Euclid's *Elements* was a collection of 13 books that built up all the mathematical knowledge known at his time. They started with definitions and postulates (axioms), and then went to propositions (theorems and constructions) to build quite a complex body of knowledge that people would use for two millennia for teaching mathematics.

This book (or collection of books) has been the most successful and influential textbook of mathematics ever published or written anywhere in the world. Perhaps you should get a copy! There are many on the internet freely available, in most languages.

Leonhard Euler 1707-1783

FIGURE 140   Euler.

13.7 LEONHARD EULER (1707–1783)

Euler was a Swiss mathematician. He went to St Petersburg as a young man after studying with the famous Johann Bernoulli (1667–1748). Euler also had the good luck of being friends with Johann's sons and eventually making it to the newly founded St Petersburg University and Academy.

One of the most prolific mathematicians of all times, his works can be seen on various websites, and many of his works have been more recently translated from Latin to vernacular languages like German and English.

He was best friends with Christian Goldbach, who also appears in this little book.

Pierre Fatou 1878 - 1929

FIGURE 141   Fatou.

13.8 PIERRE FATOU (1878–1929)

Pierre was a French mathematician and an astronomer. He studied at the École Normale in Paris (as did many of the others mentioned in this little book) and became an astronomer towards the end of his life.

He was friends with Maurice Fréchet, whom you can find in this list too.

Fatou came up with a study of iteration of special functions that would result in what would become eventually known as Julia sets.

FIGURE 142   Fermat.

13.9 PIERRE DE FERMAT (1607–1665)

Where does one begin with Fermat? He was a French mathematician who did mathematics in his spare time it seems, yet came up with so many new mathematical theorems that could easily be spread between ten or more full-time professional mathematicians.

Pierre was from Bordeaux and eventually settled in Toulouse where he became a government official and a lawyer which gave him the right to add "de" to his name.

He is most famous for his *Last Theorem*, but there are other theorems you should try to find and understand if you want to do some constructive thinking about numbers. Fermat also, with his colleague and friend Pascal, formulated, more or less the probability theory.

FIGURE 143   Feuerbach.

13.10 KARL WILHELM FEUERBACH (1800–1834)

Feuerbach was a German mathematician from Erlangen (the same place Felix Klein established his *Programme*). He came up with the nine-point circle, which appears in this book. He was also someone that other great mathematicians of this period looked up to (or rather his work); unfortunately he died quite young.

FIGURE 144   Maurice Fréchet.

13.11 RENE MAURICE FRÉCHET (1878–1973)

A French mathematician of great influence around the world, which is only now becoming apparent, Fréchet worked in different universities during his lifetime and had a huge network of students, friends, and correspondents.

He is mentioned mainly because of this network of friends, although his original contribution to mathematics is no less important. There's a great book about the correspondence between him and his friend Paul Lévy published by Springer – *Paul Lévy and Maurice Fréchet: 50 Years of Correspondence in 107 Letters* (Marc Barbut, Bernard Locker, and Laurent Mazliak).

I enjoyed reading the letters his friends sent him from all parts of the globe. Perhaps in the truly French tradition, he was a point of contact between mathematicians from Eastern and Western Europe, the United States, UK, China, and Russia.

Gersonides 1288 - 1344

FIGURE 145   Gersonides.

13.12 GERSONIDES (1288–1344)

Levi ben Gershon was born in Bagnols in the Languedoc area of France. He spent most of his life around there.

Gersonides was a philosopher and a mathematician. He only ever wrote in Hebrew, so he is not well known in cultures other than Jewish. He was so unorthodox in his time, he would have been excommunicated had most of his writing had actually been understood.

One of his minor works on mathematics is about harmonic numbers which are used in music, and this is why he is mentioned in this book. There is another mention of him here, about his work on the immortality of the soul. I would highly recommend his *The Wars of the Lord* which deals with, apart from immortality of the soul, the prophecy, human freedom, divine providence, and creation of the world. All with a mathematical slant, so hence you should see it!

FIGURE 146   Goldbach.

13.13 CHRISTIAN GOLDBACH (1690–1764)

Christian was a Prussian mathematician born in Königsberg, who came up with a Goldbach conjecture which is the reason we find him in this book. He was the best friend of Euler and a godfather to one of his children.

FIGURE 147   Grothendieck.

13.14 ALEXANDER GROTHENDIECK (1928–2014)

Grothendieck was a German-born French mathematician. His mathematics was quite revolutionary, as were his political persuasions. He was crucial in establishing the modern algebraic geometry, and in particular very abstract aspects of it.

FIGURE 148   Hamilton.

13.15 WILLIAM ROWAN HAMILTON (1805–1865)

Hamilton was an Irish mathematician who gets a mention for his Hamiltonian walks and graphs and the game he invented.

Hilbert

FIGURE 149   Hilbert.

13.16 DAVID HILBERT (1862–1943)

Hilbert was a German mathematician who listed some interesting problems, some of which are still not solved more than 120 years since they were posed.

FIGURE 150   Hypatia.

13.17 HYPATIA (C. 370–415)

A Neoplatonist philosopher and mathematics, she lived in Alexandria and was a daughter of Theon of Alexandria.

FIGURE 151   Iamblichus.

13.18 IAMBLICHUS (C. 245–C. 325)

A Greek Neoplatonist philosopher who wrote about Pythagoras and from whom we have most of the information about both Pythagoras and his followers, Pythagoreans.

Gaston Julia 1893-1978

FIGURE 152   Gaston Julia.

13.19 GASTON JULIA (1893–1978)

A French mathematician who came up with the *Julia set* which was then popularised by Benoit Mandelbrot. He and Pierre Fatou were the founders of a theory called holomorphic dynamics. It studies dynamical systems which are obtained by a process of iteration of mappings. Iteration is repeated procedure from the initial condition, using a prescribed rule.

Fujita Kagen 1765-1821

FIGURE 153 Kagen.

13.20 FUJITA SADASUKE KAGEN (1734–1821)

Kagen was a Japanese mathematician who created the first known collection of sangaku problems.

FIGURE 154   Kepler.

13.21 JOHANNES KEPLER (1571–1630)

A German mathematician and astronomer, Kepler was a key figure in the scientific revolution that took place beginning with the 17th century. He was an inspiration to Newton and provided some of the laws (laws of planetary motion) that were foundations for the theory of universal gravitation.

Felix Klein 1849-1925

FIGURE 155   Klein.

13.22 FELIX CHRISTIAN KLEIN (1849–1925)

Klein was a German mathematician who was interested in a great number of the branches of mathematics. He was a close friend and colleague of David Hilbert. Klein initiated the Erlangen programme which was aimed at classifying geometry by the basic symmetry groups.

FIGURE 156   Leibniz.

13.23 GOTTFRIED WILHELM LEIBNIZ (1646–1716)

A German mathematician and philosopher, Leibniz had invented calculus independently from Newton. His work on I Ching was quite revolutionary (although not mentioned in this book), as was his work on binary arithmetic related to it.

FIGURE 157   Paul Lévy.

13.24 PAUL PIERRE LÉVY (1886–1971)

Lévy was a French mathematician who introduced some fundamental concepts to modern mathematics with many things named after him. He was also a good friend of Maurice Fréchet.

FIGURE 158   Mandelbrot.

13.25 BENOIT B. MANDELBROT (1924–2010)

Mandelbrot was born in Poland, but being Jewish left as a child and spent most of his life in France (primarily) and the US. He is the real inventor of fractals and the one who cracked their laws and how to describe fractal properties.

Mersenne 1588-1648

FIGURE 159   Mersenne.

13.26 MARIN MERSENNE (1588–1648)

Mersenne was a French mathematician and philosopher, as well as a Minim friar. Before the invention of the internet, it is said Mersenne was the one facilitating connections between mathematicians. He is also known for his own mathematical inventions, such as Mersenne primes.

FIGURE 160   Möbius.

13.27 AUGUST FERDINAND MÖBIUS (1790–1868)

Möbius was a German mathematician and astronomer. He greatly advanced topology and there is now a strip called after him – a non-orientable surface with only one side. You must try making it!

FIGURE 161   Monge.

13.28 GASPARD MONGE (1746–1818)

Monge was a French mathematician, revolutionary, inventor of *descriptive geometry*, and founder of the École Polytechnique. The most recommended of all the meditations is the one on his principles, in the last chapter.

FIGURE 162   Nicomachus.

13.29 NICOMACHUS (c. 60–c. 120)

Nicomachus was the Neopythagorean philosopher from Gerasa, which was at the time the Roman province in Syria (now Jerash, Jordan). Like many Pythagoreans and Neopythagoreans, he was interested in numbers.

Nicole Oresme 1323-1382

FIGURE 163   Oresme.

13.30 NICOLE ORESME (1325–1382)

Oresme was a French mathematician and philosopher. He wrote influential works on mathematics, astronomy, philosophy, and theology. He was a bishop of Lisieux and a counsellor to King Charles V of France.

FIGURE 164   Pascal.

13.31 BLAISE PASCAL (1623–1662)

Pascal was a French mathematician and philosopher. He was a child prodigy and is mentioned here for his contribution to the development and the founding of the theory of probability. He came up with the first calculating machine that worked (after the abacus), called Pascaline.

Louis Poinsot 1777 - 1859

FIGURE 165   Poinsot.

13.32 LOUIS POINSOT (1777–1859)

Poinsot was a French mathematician and physicist. He studied the star polyhedra of Kepler, and they are now called after both of them. He established geometrical mechanics, the science of physical forces in mechanics that can be described in geometrical terms.

FIGURE 166   Pythagoras.

13.33 PYTHAGORAS OF SAMOS (c. 570 BCE–c. 495 BCE)

Pythagoras was a Greek philosopher and mathematician, and a founder of the Pythagorean sect. Most of the work he and his followers did is a bit veiled in myth and mystery. But apart from the myth and mystery, we also have from them the foundations of the number theory.

Bertrand Russell 1872–1970

FIGURE 167   Russell.

13.34 BERTRAND RUSSELL (LATER LORD RUSSELL, 3RD EARL RUSSELL, 1872–1970)

Russell was a British mathematician, philosopher, and eventually a public intellectual. He also won a Nobel Prize in literature without writing a novel (he did after he got the prize) and was mainly interested in the foundations of mathematics. We have to thank him for the Russell paradox.

FIGURE 168   Theon of Smyrna.

13.35 THEON OF SMYRNA (FLOURISHED AROUND 100)

Theon was a Greek mathematician who worked on number theory. He wrote on prime numbers, perfect numbers, abundant numbers, and other numbers that have characteristics that make them interesting more than any other ordinary numbers. He wrote also on the *music of the spheres*, which is what he is mentioned here for.

FIGURE 169   Zermelo.

13.36 ERNST ZERMELO (1871–1953)

Zermelo was a German mathematician who worked on the foundations of mathematics. We mention him here for his work on axiomatic set theory and his famous *axiom of choice*.

Post Scriptum
and All That

Don't forget, this book was done to hopefully inspire you to enjoy some mathematics. If you don't like some topic or area of mathematics, skip the pages until you can get to a place where you will enjoy thinking about things and drawing, writing, or just imagining them. You are meant to go back and forth, although you may prefer to just stick with the order and go through the book as it is.